THE FINE ART OF THE
TIN CAN

■ ■ ■

TECHNIQUES AND INSPIRAT...

PHOTOS & TEXT BY

BOBBY HANSSON

Lark Books

Published by Lark Books
50 College Street
Asheville, North Carolina, U.S.A. 28801

© 1996 by Bobby Hansson

Editor: Chris Rich
Art Direction: Kathleen Holmes
Production: Elaine Thompson

ISBN 1-887374-02-7

Library of Congress Cataloging-in-Publication Data
Hansson, Bobby.
    The fine art of the tin can : techniques and inspirations / photos
and text by Bobby Hansson.
       p.   cm.
    Includes bibliographical references and index.
    ISBN 1-887374-02-7
    1. Tinsmithing.   2. Tin cans.   I. Title
TT266.H35   1996
745.56--dc20
                           96-4039
                           CIP

10 9 8 7 6 5 4 3 2 1

Distributed in the U.S. by Sterling Publishing, 387 Park Ave. South,
  New York, NY 10016; 1-800-367-9692

Distributed in Canada by Sterling Publishing, c/o Canadian Manda Group,
  One Atlantic Avenue, Suite 105, Toronto, Ontario, Canada M6K 3E7

Distributed in Great Britain and Europe by Cassell PLC,
  Wellington House, 125 Strand, London, England WC2R OBB

Distributed in Australia by Capricorn Link (Australia) Pty Ltd.,
  P.O. Box 6651, Baulkham Hills Business Centre, NSW 2153, Australia

Every effort has been made to ensure that all information in this book is
accurate. However, due to differing conditions, tools, and individual
skills, the publisher cannot be responsible for any injuries, losses, or other
damages which may result from the use of the information in this book.

Printed in Hong Kong

Cover, clockwise from upper left:
Holy Ghost Building #22, Christina Shmigel
Bird Brooch, Professor Robert Ebendorf
Truck, Bobby Hansson
Dreamin' of the Soul Roundup, John Grant
Grand Opening Purse for Zette, Bobby Hansson
Shaping a Handle, Bobby Hansson
Tin Can Alley Sign, Sue Eyet
Captain's Lantern, Richard Haddick
(Center) *Can*, Bobby Hansson
Back Cover: Laurie Flannery

■ *Man Holding the World*
John J. Grant

A Letter from the Author
to His Editor and Art Director

Dear Chris and Kathy:

Well, here it is—an attempt to acknowledge and thank my mentors and to pass on some help to the next generation. I hope that The Fine Art of the Tin Can *will sound as if a real person wrote it. I'm very serious about art and craft, but I believe there's room in life—and in arts and crafts—for fantasy, humor, and even silliness. I'd like the book to reflect this belief.*

I want the reader to feel empowered by every page. I want to share as much as possible without being condescending. I don't want to answer every question, but I do want to show which paths to follow to find answers.

Photos are worth a bunch of words; I'd like to use as many as possible. I'm submitting more than three hundred.

I've done my best. I hope it isn't too much of a mess. Good luck—and thanks.

Bobby

INTRODUCTION

*From my earliest youth, I have always found pleasure in giving
life to old objects that have lost their usefulness, in delivering
them from obscurity and transforming them into objects of art.*

—*Jose de Creeft*

■ *The Eyes of Others No. 79* (1992), **Tony Berlant**
17" x 17" (43.2 x 43.2 cm)

I'VE BEEN FASCINATED by tin cans all my life. I remember watching my Mom do what she called canning—cooking vegetables from our Victory Garden and sealing them in hot glass jars—but I could never figure out how food got inside sealed tin cans. Opening up a can and looking down at those tiny peas floating in that green water made cans and canning seem like magic. Whenever I smell a freshly opened can of peas, I remember that boyhood kitchen.

World War II was on and so was the great scrap drive. After Mom poured the peas into a pan, she'd rinse the can, remove the bottom, and place the top and bottom inside the can's body before stepping on the can to flatten it. Sometimes, as a treat, I was allowed to work the can opener on the bottoms and to do the flattening. I was helping to win the war.

I remember my Dad making an *Uncle Sam Hat Bank* by soldering a soup can onto a tomato-juice can lid. He used a knife to punch a slot in the top. Then he hammered a quarter through the slot to size the hole and to give my savings a start. When I filled the bank up, I got to cut it open and take the money to school to buy a bond.

After we won the war, I used to thread long loops of twine through the church-key holes in tomato-juice cans and then pull the cans up tight against the soles of my sneakers so I could walk in puddles without getting wet feet. When I went fishing, I kept my worms safe in a tin can. Tuna cans made swell headlights for my soap-box race cars. My high-school pal, Buckley, and

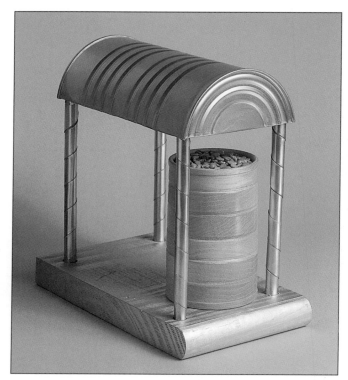

■ *Pecan Can of Peas with Tin Can Canopy,* Bobby Hansson
A board from a pecan tree was sanded flat, cut into squares, glued in a stack, and turned on a lathe. The "can" that resulted was filled with split peas and covered with a split can awning.

■ *Uncle Sam Hat Bank,* Bobby Hansson

■ Professor David Driesbach

■ *Knife and Sheath,* Bobby Hansson
I made the knife (from an old file) and the sheath (from an olive-oil can) during the time I spent with de Creeft.

I made diving helmets out of five-gallon cans, complete with plastic windows and lead-filled orange-juice can weights. We relied on bicycle pumps for air. To this day, I keep my pencils and crayons standing in a row of cans on my desk.

While I was in college, I met a man who changed my life—Professor David Driesbach. An artist, the professor taught me to draw and encouraged me to make constructions out of found objects. For the first time in my life, I knew what I wanted to be when I grew up—an artist.

My friend, "Racer" Wiley, showed me how to weld, and soon my parents' backyard was heaped with tons of rusty "treasures." I worked as a photographer's assistant and sculpted on my days off.

During this period, I had the chance to photograph some stone sculptures by an 80-year-old artist named Jose de Creeft, and I started to spend time in his studio, watching him work.

"Someone ought to make a film of Jose carving stone," I said to friends one day.

"Why don't you?" they replied.

I quit my job, bought a 16mm camera, and moved into a small shed in de Creeft's backyard.

Jose, who was born in Spain in 1884, left school when he was eight or nine. By the time he was thirteen, he was working as a sculptor's helper and was soon creating his own pieces. With the exception of a day off here and there to draw or paint, Jose worked at his sculpture seven days a week until his death at 98.

Jose never wasted time or materials. Like his friend, Alexander Calder, the famous American sculptor who invented the mobile, he transformed found pieces of wire and tin into jewelry for lady friends and toys for children, as well as into art.

During the months I lived with Jose, I learned how to carve stone and make tools in his blacksmith shop. More importantly, I saw a way of life I wanted to follow.

■ Alexander Calder and Jose de Creeft

■ *Le Picador,* Jose de Creeft
Jose was deeply moved by the plight of the horses martyred in bullfights. Using stovepipes, oil cans, bicycle parts, and other odds and ends, he created this blind-folded nag, which stumbles along, spilling inner tubes and wires from its gored belly as the proud picador sits, ramrod straight, his curtain-rod lance at the ready. Ramshackle materials contribute to the tragic eloquence of this revolutionary work, created in Paris in 1925.

CHAPTER ONE:

TIN CANS, MIXED MEDIA, AND TOOLS

■ *Dreamin' of the Soul Roundup,* John J. Grant
John is an artist from California who takes tin collage to its farthest limits. He claims he can only draw with tin snips.

MOST DESIGN approaches start with the function of the object to be made. If we were designing a wheelbarrow, for example, we'd consider its uses and choose shapes and materials that would best serve them. In this book, we'll be working backwards. We'll start with the tin can as a material and will design objects that take advantage of and are in harmony with its unique properties.

We'll be using cans as cans, taking advantage of their shape, construction, and decoration. We'll also "deconstruct" cans for their tin, which we'll use just as a tinsmith uses flat sheets of brand new tin. Our tin cans will be free for the taking, of course. In the United States, hundreds of thousands of cans are used every year and then discarded, but their availability won't be why we're using them. The projects we'll be exploring will be made from cans because there's no material better suited for them.

TIN CANS AREN'T MADE FROM TIN

The old British term "tinned can," derived from the term "tinned canister," is more accurate than our common term "tin can." Tin cans are made of tinned steel—steel plated with tin. The steel provides strength and economy, and the tin resists rust and corrosion.

The process of tinning iron was invented in Bohemia at the beginning of the sixteenth century. The earliest tin cans were made from tinned iron so thick and strong that soldiers often resorted to bayonets and hammers to open them. The can opener wasn't invented for

another 50 years, by which time cans were made of slimmer stuff.

USING IT ALL

I once heard of a farmer who paid a visit to a newfangled meat-packing plant in Chicago, where he was treated to the full fifty-cent tour. The tour's highlights included the meat-cutting facilities, the hide-processing and bristle-brush areas, the bone-button works, and the sausage department. Back down on the farm, he regaled his cronies with as many facts as he could recollect and finished by declaring, "I'll wager those fellows had a use for every part of that pig excepting the squeal."

In *The Fine Art of the Tin Can*, I've attempted to cover every artistic use for the tin can, including the squeal. If you doubt this, turn to page 102 and take a look at Craig Nutt's Folgerphone. You'll even find the shadow of a crushed can, captured by a photocopy machine (see page 131).

COMBINING MATERIALS

Using tin cans in combination with other recycled materials adds greatly to the number of possible projects. Strength, visual interest, humor, and convenience all play a part in deciding to use mixed materials.

Although tin-plated steel is quite strong, reinforcing it with other materials is sometimes a good idea. The most common of these materials is stiff wire. If you bend a channel along the edge of the tin, lay a piece of stiff wire in the channel, and then bend the tin to form a seam around the wire, the strengthened edge will hold its shape better. By closing the seam with solder, you can make the edge even stronger.

This wiring method is often used for making handles. Peter Ross's reflector oven on pages 66-68 offers a fine example of this technique. Large items such as this oven often include wire around the rims. Richard Haddick's sconces include

■ *Sconces*, Richard Haddick

wire in the arm sections. An interesting and attractive variation is to lace the tin onto a wire frame; the knife, fork, and spoon on page 55 were made by using this technique. To construct these tableware pieces, a very thin wire was looped through tiny holes and around the individual frames.

Wood is another material that's often used in combination with tin cans. The wooden base of the green and white bus shown on page 86 makes it a more durable toy, stabilizes the structure, and also adds the weight necessary to help the bus roll along smoothly. Randall Cleaver's clock (see page 56) includes a large wooden block inside the base. An object as tall and narrow as this would otherwise be so top-heavy that it might tip over. The toy tank pictured on page 35 has wooden sides; these add weight, reinforce the structure, and provide a convenient base to which the bottle-cap gears can be attached.

Many of Harvey Crabclaw's frames incorporate a wood or plywood backing. One of Harvey's main concerns is keeping the art he's framing flat and smooth; gluing it to wood is a good way to do this. Like John Grant (see page 11), Harvey often uses small brads to fasten his cut-out tin shapes to a board. The brads hold the tin securely in place, and their heads become part of the art — punctuation marks of a sort.

Robert Dancik's bracelets (see page 31) are covered with pantyhose, which provide texture, color, and a comfortable means of holding them onto the wrist.

Many jewelers use tin cans as a decorative element to add color, texture, and content to their silver jewelry. The tin can buckle on page 25 wouldn't be strong enough to hold up a fat man's pants by itself, so it's fastened to a welded steel buckle with copper rivets that almost match the painted color of the can.

■ *M.M.*, Harvey Crabclaw
Here's Harvey's valentine to the queen of the silver screen—a frame which he made from a tin candy container and decorated with a steel letter stamp.

TOOLS OF THE TRADE

Tin may be cut with shears, scissors, knives, chisels, saws, or torches. Keep in mind that because tin is thin and stiff, it can also cut back, easily slicing or jabbing through cloth, leather, rubber, and/or your skin. If you work with tin long enough, you'll probably shed a few drops of blood, but observing a few basic safety rules and keeping your wits about you will make working with cans no more dangerous than raising kittens. Develop safe working habits!

—**It's always wise to wear safety glasses when you work with tin, especially when you're using power tools.**

—**Every time you cut a piece of metal, put the waste part into a trash bucket or a "big-enough-to-save" box. Cut metal can't hurt you if you don't leave it lying around.**

—**Remove burrs from cut edges right away.**

—**Keep your work area tidy. When you're deeply involved in what you're doing, it's sometimes tempting to put off cleaning up until the next day, but keeping your work area orderly as you go will prevent accidents.**

—**Alternate tasks as you work. Repetitive hand motions can cause hand cramps and carpal tunnel syndrome.**

The pinching-class tools shown above are all handy for tin work. Pictured to the left are square-jawed needle-nose pliers, a pair of glass-breaking pliers, a homemade tin bender (built by welding steel-ruler sections onto the jaws of a regular pair of pliers) and a pair of tinsmith's bending pliers. These four pliers are used for grasping the tin and making straight bends.

At the top of the photo is a pair of modified lineman's pliers—with rounded jaws—for bending smooth curves. Below these modified pliers are a typical pair of lineman's pliers, for cutting wire and can rims and for general pinching. The needle-nose pliers on the right are for bending metal in tight quarters.

Some other tools you'll find helpful are shown below. The can

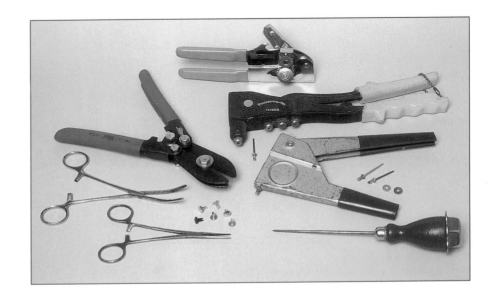

opener needs no description. (A wall-mounted model is even better!) The yellow- and blue-handled tools are pop rivet guns, shown with pop rivets and optional washers. The ice pick with a 1/8" (3 mm) shaft is used to make rivet holes, and the surgical clamps are ideal for clamping thin tin while soldering. The red-handled tool will crimp wavy edges into sheet metal. Split rivets (see page 50) are shown to the right of the surgical clamps.

You'll want to have a wooden or rawhide mallet, a ball-peen hammer, a few sturdy blunt and sharp knives, and some hardwood dowels or metal rods. A bench vise will also come in handy.

■ *Truck,* **Bobby Hansson**

BOBBY HANSSON

As a teenager, I tried several tin can projects, some of which worked better than others. "The Boy Scout Shower" involved punching holes in the bottom of a #10 tin can and hanging the water-filled can from an overhanging tree limb so I could take a refreshing outdoor shower. Bad idea. Pouring water directly onto myself from an unpunctured tin can would have worked much better.

For frying "Baloney and Eggs with a Metallic Flavor"—a favorite recipe of mine—the big upside-down tin can that I balanced on rocks was a better idea. So was protecting camping provisions from mice by hanging the food from a tree branch and threading a tuna can onto the rope; the mice couldn't shimmy down past the can to my food. Raccoons, however, laughed at this attempt to stop their pilfering.

The Chelminski brothers once tried manifold cookery, which seemed to be a brilliant idea at the time. Bound for a pre-dawn ski outing in Vermont in subzero weather, they fashioned a coat-hanger cradle to hold a large can of pork and beans against the exhaust pipe of their four-wheel-drive Willys. A nail hole in the top of the can kept the can from bursting as it heated.

"By the time we get to the slopes, those beans will be piping hot," one brother declared. Unhappily, the frigid New England wind had as much effect on the far side of the beans as the engine had on its half, producing a concoction that was half burnt and half frozen. Undaunted, the famished lads hatched plans for a revolutionary can turner as they ate every last morsel.

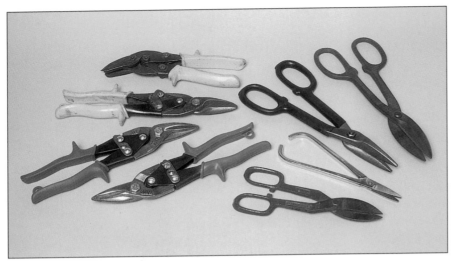

Pictured above are some of the tin snips and shears commonly used for cutting tin. On the left are four pairs of aviation-style snips, the compound hinges of which give them increased leverage for easier cutting. Note that the jaws on each pair are slightly different. One pair is for cutting straight lines or slight curves; one is for cutting right-hand curves; another is for cutting left-hand curves; and the pair at the top are called nibblers. (Turn to page 85 to see these being used to make aerosol-can wheels for a truck.) The snips on the right are basic variations of the standard types of snips that have been used for hundreds of years.

Although it's certainly possible to create tin can art without using a soldering iron, owning one will open up a huge range of possibilities. In the photo below, you'll see three typical electric irons, as well as an antique iron which is heated over a fire. An electric soldering iron with a pointed tip is a good iron for beginners. You'll also need a roll of solder and some paste flux or liquid flux.

Most of these irons come without switches; they heat up when you plug them in. Don't forget to unplug your iron as soon as you're finished using it. I plug mine into a heavy-duty multiple outlet with a switch. I also have a small work light plugged into the same strip; the light reminds me when I've forgotten to turn off the iron.

The tips of soldering irons get very hot; they'll burn you if you touch them and can start fires if left in contact with wood, paper, or cloth. Work on a brick, tile, or metal surface and don't keep anything flammable nearby.

Soldering fumes are toxic, so work in a well-ventilated area. Most solders contain lead. If you'd like to cut down on the danger of working with this substance, buy low lead or lead-free solder, and be sure you use lead-free solder when you're making drinking cups or food containers.

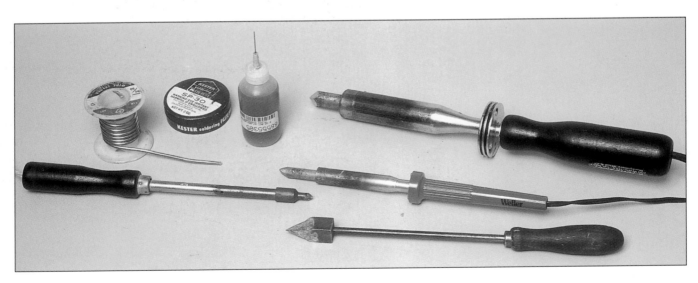

CHAPTER TWO:

WORKING WITH SINGLE TIN CANS

■ *Auto Drip II (1994),* Daniel Eaves
Here's a working coffee maker and pot with coffee-can "skins."

OUR FIRST project—the intact, unopened can as a paperweight—needs no further elucidation or instruction.

Our second project—a pencil holder made from an empty can with one end removed—needs only this cautionary note: Mash down any burrs on the inner surface of the open end. To do this, just set the open can on its side and press the back of a spoon against the burrs on the inner rim, rolling the can to reach each rough spot.

Our third project—the *Tuna-Can Cup*—is only slightly more complex. With apologies to readers who have already figured out some of what follows, let's start by grabbing one of the magnets off the refrigerator door and going shopping.

For the project you're about to make, you'll need a 6-1/2-ounce (192 ml) tinned steel—not aluminum—can. It should be 1-5/8" (4.1 cm) in height and 3-1/4" (8.2 cm) in diameter and both its top and bottom should be seamed.

The tuna-fish section of a well-stocked market will offer at least two types of tuna cans. One type has a seamed lid at both ends. The other type has a seamed top and a narrower, formed bottom; these are the cans that fit together so well when they're stacked. Some formed-bottom cans are made of aluminum; you'll find plenty of them if you wander into the cat-food section.

As I stop at a market aisle to check out cans, I like to pretend I'm reading the labels for cholesterol information. What I'm actually doing is surreptitiously touching the refrigerator magnet to the bottom of a

■ *Cups and Candlesticks*
Bobby Hansson
This motley group shows a range of one-can design possibilities. The transmission-fluid drinking cup is intended as a joke.

can. If the magnet sticks, the can is steel. If it doesn't, the can is aluminum. Always test the bottom of the can, not its top or sidewall. The tops of some aluminum cans are steel, and the paper labels on the sides of some cans will thwart a weak magnet.

TUNA-CAN CUP

When you're back home and ready to get started, open the can with an ordinary can opener, but instead of removing the whole lid, leave about 1" (2.5 cm) uncut. Use a blunt knife to pry up the lid until it's perpendicular to the body. Be careful; the lid

is sharp! Empty the can and wash it with soapy water to get rid of the tuna aroma. Examine the inside of the rim and mash down any sharp burrs.

Now let's turn the lid into a handle. First, you'll need to shape the circular lid into a rectangle by folding over both edges as if you were folding a letter before putting it into an envelope. Making these folds will create smooth rounded edges on the handle.

An easy way to get a perfectly straight fold line is to clamp the lid into the jaws of a vise, as shown in the photo at the top of this page. Clamping it between two straight-edges will also work. After you've made about half of the fold, remove the can from the vise or clamps; the crease in the metal will remain straight as you complete the fold. To tighten the fold, tap its edge lightly with a wooden or rawhide mallet.

■ *Kitchen Projects,* **Bobby Hansson**
The scoops have intact lids at one end; their handles are made from the sidewalls of the cans. The sauce pitcher has an added spout. The cookie cutter was sliced from the top of a lid-less can; a handle was attached across its top.

Repeat to fold the other side of the lid. You should end up with a parallel-edged tab.

To shape the tab into a handle, slowly wrap it around a wooden or metal rod such as a broomstick (see the photo at right). If you notice crimps developing, use a mallet to tap them down as soon as they appear. To get a smooth, even loop, bend the tab gently and hammer out any wrinkles right away.

The cup handle that you're shaping will be a little uneven where the bent portions overlap. If you want a perfectly smooth handle, use tin snips to trim some metal from the lid before you bend the two edges inward (see the can in the center of the photo at right).

When your semicircular handle is almost complete, clamp 1/4" (6 mm) to 3/8" (1 cm) of its end in a vise, or grip the end with pliers, and bend the metal to a 90° angle. This section will rest flat against the side of the

■ *Tableware,* Bobby Hansson
If matching tableware appeals to you, study these variations on a theme. Several handles are soldered at both ends, while others are made from sidewall that has been looped down and soldered at the bottom. The handle sticking up shows how a sidewall handle can be extended by folding in and soldering on another section—in this case, the label from the can.

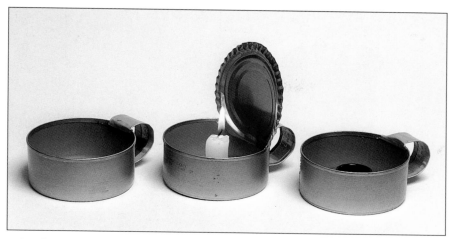

■ *Candlestick Variations*, Bobby Hansson
These candlesticks are all based on slight variations of the cup design. One contains a bottle cap to hold the candle. A bent lid creates a reflector on another.

can, where you'll fasten it by soldering it or inserting a rivet. To prepare for riveting, use a nail, drill, or ice pick to make a 1/8"-diameter (3 mm) hole in the handle. For riveting and soldering instructions, turn to pages 49-50. With a little soldering practice and some lead-free solder, you can turn your cup into a waterproof measuring or drinking cup.

It's fun to see how many variations you can create from a can with a connected lid. An act as simple as sandpapering the surface, for example, will add an interesting texture. As you'll see in the upcoming pages, short wide cans work best for these projects, because the proportionately large lids allow for more variations.

By using the techniques you've already learned and making a few modifications, many of the items pictured are yours for the making. Let the others serve as inspirations!

■ *Menorah*, Bobby Hansson
I made this Menorah—the ceremonial candle holder that Jewish people use during the celebration of Hanukkah—from a single tin can. The top of the can was turned into the base, and strips from the sidewalls of the can became the nine candleholders. If your faith is strong and pure, a Menorah made from a tin can is just as meaningful as one made of gold.

■ *Holder for Hot Glasses,* **Bobby Hansson**
I was inspired by Jerry's Hovanec's dipper (shown to the left) to make a couple of tin glass-holders like the ones found in soda fountains during the 1950s. Finding glasses to fit the cans was hard, but not as hard as blowing hot glass into a can, and making these glass/holder combinations didn't burn off the paint, either.

■ *The Hunt for the Big Dipper,* **Ruthann and Jerry Hovanec**
My old pal, Jerry Hovanec—a glass-blower and dedicated joke player—has pushed the limits of the connected-lid design about as far as possible with this dipper. When he found a Hunt's tomato can, "The Hunt for the Big Dipper" became the working title for this piece.

With the help of his wife, Ruthann, he cut both the top and bottom of the can partially open and made the handle by folding the lids to meet in the middle. Ruthann used a small sharp nail to punch the words "big dipper" in the side and a larger nail to punch in a diagram of the stars.

Then all Jerry had to do was form the dipper by blowing some blue glass into the can. If you have to ask how he did this, you probably aren't ready to try it. I know I'm not, so I won't include instructions here. As you may have noticed, the molten glass burned the paint off the can, removing part of the joke, but the burnt surface struck Jerry as ghostly and beautiful, so he was as pleased as he was disappointed.

■ *Klassic Katch,* **Karen Lasley**
Karen, whose company is called Klassic Spirits, gets her son Todd to crush empty aluminum cans, hundreds at a time, by running over them with his car. As she studies each crushed can, images appear to her, and she paints the cans to turn those images into art.

■ *Juicer (1991),* Daniel Eaves
The artist refurbished this working juicer.

DANIEL EAVES

I first started to use tin cans in my artwork to create serving pieces. The construction techniques were similar to those I had learned from my father, a former sheet-metal worker with the railroad. I managed to find some old tinsmithing tools and started to work with empty coffee cans. Before long, I'd completed my first "coffee cup."

I was entirely fascinated by the idea of having an object's material and decoration reflect its use. This was a new area, one in which my art could expand, and it led to complete coffee services, functioning percolators, auto-drip coffeemakers, orange juicers, tomato-sauce pans, and other appliances.

The tin cans and their decoration enable me to address contemporary art and design in today's household. The use of tin cans provides a lively new dimension to appliances and provides an escape from the monotony of everyday household design. Their use enchants and, if we smile or laugh at their construction, decoration, and function, then so much the better.

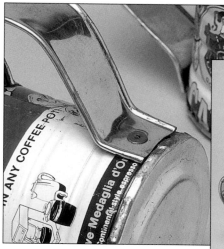

■ *Gold Medal Coffee Cups: 1993*
Daniel Eaves
These cups are examples of a more sophisticated technique, in which the cans are reduced in diameter as well as in height. The bottoms were removed, cut down, and reattached. The cups' upper rims are reinforced with wire.

■ *Coffee Mug,* Elizabeth Brim
A blacksmith at The Penland School in North Carolina, Elizabeth forged the handle for this mug from a steel bar and then welded it onto a selection of radiator hose clamps. To wash the mug in a dishwasher without getting the handle rusty, she just loosens the clamps with a screwdriver, and the tin can slips right out.

■ *Coffee Pot,* Jim Wallace
While Jim was teaching a blacksmithing course at The Penland School, he decided that the shop should have a large pot of coffee available at all times. He requisitioned a couple of #10 tin cans and built this pot, which was designed to be used on the coal fire of the forge. To keep flying cinders out of the brew, Jim made the lid overlap the spout. He hammered out the handle and lid decoration from mild steel and brazed the whole construction together. The brazing is rough, Jim explains, because he had to improvise by using a cutting torch, which has a very wide flame, for the job.

■ *Tinderboxes*
Right: Richard Haddick
Left: Bobby Hansson
Richard made the tinderbox shown in the background as a replica of a historical work. The fire-starting equipment—a piece of flint, a steel striker, a charred linen cloth, and some shredded hemp—is stored inside the box, which has a candle on top to keep the flame going.

I found the Jack Daniels square can with a lid and, from it, got the idea of making a tinderbox. I drew a circle in the middle of the lid by tracing around a penny. Then I made a star-shaped hole in the circle by pushing a sharp knife down through it several times, in much the same way as you'd cut a pizza. Using a dowel, I pushed the wedges of cut tin upward. I then made a tin cylinder about 1" (2.5 cm) long that would fit in the hole and support the candle. I added a handle, and the project was finished.

I often use this tinderbox to light my forge when I give Colonial blacksmithing demonstrations. I can't resist telling folks that when I want to get lit, I rely on my whiskey box.

■ *Book,* Lucille Beards

Like her dear friend, Louise Nevelson, Lucille began making artwork from found objects many years ago. Nevelson died in 1988, but Lucille, now 84, still makes collages and sculptures. The piece shown here is a book created for a poem entitled "Historic Havre de Grace," written by her grandson, James. (Lucille lives in this small town, which she treasures.)

She made the book by gluing the lines of verse and images of the city onto crushed cans that she'd picked up on the streets. The book is punched and is bound with a steel notebook ring. It's pictured here both open and closed.

■ *Buckle,* Bobby Hansson

I found this crushed can on the street and was amused that it said "Crush" on it. The can's texture, color, and shape added to its appeal, so I took it home, washed it off, and used copper rivets to attach it to a steel belt buckle. I thought the rivets looked good with the orange printing. My son likes to borrow the buckle when he goes slam dancing.

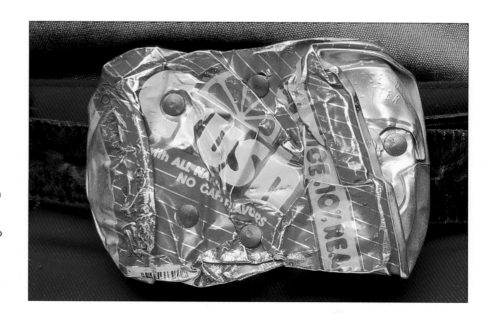

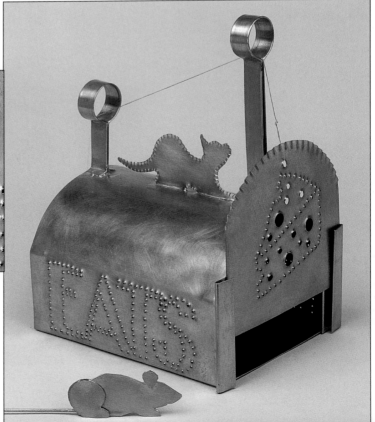

■ *Mousetrap,* Rob Hudson

As a master bladesmith, Rob makes some very scary-looking knives, so I was especially amused by this mousetrap in his shop—a "catch-'em-but-don't-hurt-'em" tomato-can variety instead of the guillotine-like device I would have expected. Rob generously built this one for me and included a drawing with instructions. We decided to add some lettering from another can, so the label now reads "Hunt's tall natural mice." I've tested this project, by the way; it really works well.

■ *Instructions for Mousetrap,* Rob Hudson

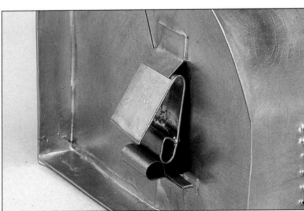

■ *Mousetrap,* Lindsay Janet Roberts

I bought this mousetrap from the Whip-Poor-Will Studio in Logan, Ohio. It's made from regular sheet tin, but tin can metal would be a fine substitute. The trap comes with a tin mouse so you can demonstrate how it works. Take a look at the trigger mechanism, shown in the photo above.

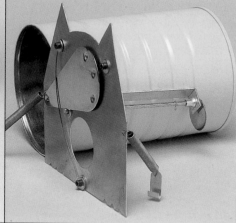

■ *Hamster Trap,* **Chris Darway**

A well-known jeweler, Chris creates hard-edged, mechanical, very elegant work. Late one night, he was informed that the family hamster had escaped and had to be recaptured before the family cat converted him into breakfast. Chris found a coffee can in his studio, fired up the hot-glue gun, gathered together some acrylic plastic sheets, iron wire, and a few other materials, and the result was a gravity-drop, pivot-door trap. The bait was a glob of peanut butter on the round disk inside the trap.

At 3 a.m. the next morning, Chris heard a click. The hamster was safe in the can. Everyone was impressed, so Chris decided to make traps more aesthetically pleasing and to start selling them.

The photo on the left shows the set trap, and the one on the right shows the same trap opened. When the peanut butter is nibbled, the disk slides back, and the rod to which it's attached slips out from under the door. Gravity causes the door to pivot closed. To make up for having thwarted its breakfast plans, Chris created a portrait of his cat on the front of the trap.

■ *Bird House,* **Sue Eyet**
This teeny bird house includes a custom-made copper roof.

MARCIA WILSON

I fired up my oxyacetylene mini-torch, which I use to solder copper jewelry. Its tiny blue and orange flame melted the tin very quickly; I found the experience quite exhila-rating. I didn't plan ahead of time. I just hit the cans with the torch and let the figures come out. I thought of dancers by Matisse, Pennsylvania Dutch tulips, and vicious dogs.

It was an intoxicating feeling, sitting in my backyard as the sun went down, using fire to cut shapes in the cans. I remember thinking it was just about the best thing in the world a person could be doing on an August evening.

My boyfriend the engineer said liners would help make the designs stand out on the cans, so I returned to the recycling center to look for straight-sided white plastic containers. Obesity-pill bottles fit snugly into the vegetable cans, and shampoo bottles slipped right into the Mandarin orange cans. I trimmed the bottle tops with a sharp knife.

I thought about painting these cans, and I thought about rusting them, and then decided to leave them alone.

■ *Candleholders,* Marcia Wilson
These torch-cut cans with plastic liners seem to be perfect candleholders, although frosted-glass liners would probably be safer. Notice how the design is connected to the top and bottom rims at regular intervals. Marcia didn't make any preliminary sketches, but "drew" directly on the metal with her torch. The melted outlines work very well with the shapes she has designed.

■ *Flame-Cut and Painted Cans*
Marcia Wilson
This ferocious pooch isn't having much luck scaring the cat off the square cookie-can lid they share. Wilson owns several bad dogs and one inscrutable cat.

■ *Flame-Cut and Painted Cans*
Marcia Wilson
Color is very important to Marcia; she couldn't resist painting some of these cans with regular household enamel paint. The "Ring Around the Rosie" design is very cute, especially with its stripes, which follow the ribs of the can.

■ *Flame-Cut and Painted Cans*
Marcia Wilson
These perky cans, a colorful group of ladies and flowers, could be used to hold any number of items, from pencils to dried flowers, and would brighten up even the gloomiest of corners.

■ *Flame-Cut and Painted Cans*
Marcia Wilson
The hats in this group were added on with a pop rivet tool. For a smoother look, Marcia hammered down the rivets even more.

■ *Water Wagon,* Artist unknown
Renee Habert and Jim Stonebraker bought this tin can water wagon in Mexico.

■ *Sanding Cylinder,* Robert Seetin
Late one night, Robert, who is an extremely creative design student, needed a tool to sand the inside of a hole on one of his projects. Even in Manhattan, where he lives, all the stores were closed, so Robert built this sanding cylinder from a coffee can. The cylinder includes a bolt so that it can be inserted into an electric drill.

■ *Tea Time,* Bobby Hansson
The next time someone invites you over for tea, make a nice quick present to bring along. Buy an inexpensive quartz clockworks, stab a hole for the spindle in a tin can lid, and connect the hands to the protruding spindle.

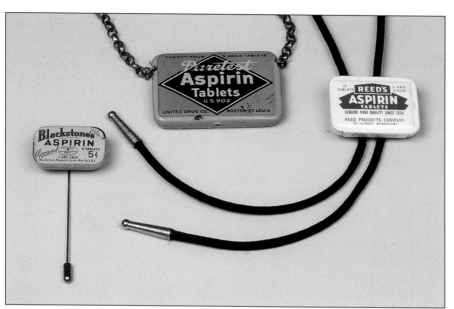

■ *Jewelry* (above and right), Janet Cooper
Janet made these charming pieces of jewelry by adding some hardware to antique cans. The aspirin boxes contain real aspirin, which could come in very handy.

■ *Bracelets,* Robert Dancik
To make these bracelets, the artist used flavored-coffee cans, which he covered with pantyhose. The one with tan hose is for daytime wear and the one with black textured hose is for evening wear. As well as being decorative, the fabric provides a soft, snug fit.

■ *Crushed and Painted Cans* (left)
Tina Chisena
When Tina, who is a scientist as well as an artist, needs a crushed can, she not only crushes it herself, but also figures out why it ends up looking the way it does so she can crush other cans to look just like it. The cans shown on the left side of this page are the results of just a few of the experiments she's conducted by standing a can perfectly straight under a ten-ton press and very slowly closing the jaws. Notice how the cans crumple in a symmetrical pattern. Very neat. These cans were painted to dramatize their wrinkles.

■ *Untitled* (above)
Edmundo de Marchena Hernandez
A metalworking student, Edmundo opened this can by cutting a round hole in its side. He then riveted a spring-wound motor to the top. A plastic soldier, heart bleeding and holding a flag, spins behind the net-covered window. Trying to stop loving someone is harder than stopping a war.

CHAPTER THREE:

WORKING WITH MORE THAN ONE CAN

■ *Briefcase*
Bobby Hansson

I DO sometimes buy products in the supermarket as much for their packaging as for their contents; printed cans are especially tempting. As you begin to work with more than one can at a time, however, you'll obviously want to find less expensive and more environmentally friendly sources for raw material.

Dumpster diving (looking through trash bins in search of raw materials) is an activity more attractive to some folks than it is to others. Wally Yater, who made the nesting bells on page 48, told me he would much rather go dumpster diving than wait in line at a hardware store. I have to agree with him, but if the thought of rooting through trash for treasure inspires only revulsion, don't despair. You do have other resources!

My friends and neighbors are happy to save interesting tins for me. My wife found a nearby tin can factory, where I was able to negotiate with the owner for scrap tin. School cafeterias can be a source of big containers, and ethnic restaurants are often willing to save cans, some of which, with their unfamiliar words and imagery, are very exotic.

Several of the artisans whose work appears in this book interrupted the process of tin can manufacture to obtain printed sheets of tin before these sheets were slit and rolled into cans. Tony Berlant (see page 127), who has made tin murals as long as 42' (12.8 m), gets misprinted and smeared sheets from a helpful tin printer.

If you choose to go dumpster diving, I'd like to offer two words of caution: Be careful! Mixed in with the goodies you may find are sharp objects, filthy objects, and even poisonous objects. I keep an old broomstick, with a hook on one end, in my vehicle, along with an assortment of plastic bags. Climbing onto or into dumpsters is asking for trouble, so I use the stick—not my hands—to do my rummaging. It lets me hook things that are out of reach. The bags are imperative when I hit a real bonanza. I also bring along a pair of heavy-duty gloves for handling what my stick catches.

Official recycling bins are an obvious place to find cans, but there are two drawbacks that come hand in hand with this source. Bees swarm all over the bins, especially in the fall. Self-righteous do-gooders are present year round.

Last summer, after dropping off my recyclables, I was doing a little

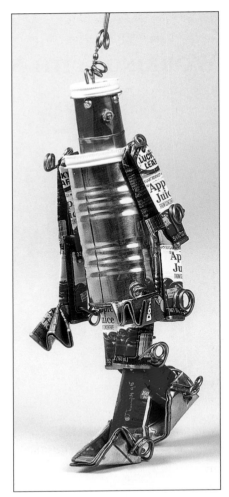

■ *Tin Man,* **Jack Trimper**
This robotic tin can man is walking, with a little help from a human hand not shown in this photo.

■ *Sign,* **Sue Eyet**

prospecting when one of the men who oversees the operation came up to me and declared, "You'll have to stop now. A man just came to my office and told me to call the police because someone was out here stealing cans."

He said that it wasn't legal for me to swap my trash for other people's, and he asked me to stop, at least until the citizen watchdog left. He personally thought that what I was doing was a better use for the material than melting it, but not all custodians at dump sites are as friendly as he was.

Always ask permission to salvage cans; explain what you want and why. Usually, the guardians of recycling centers are more interested in hanging on to valuable aluminum cans than they are in keeping you from taking tin cans, but a dog-in-the-manger attitude is common at many recycling centers, so be polite and be careful. Arlo Guthrie made a lot of money singing about being arrested for garbage, but I'm not as cute as he is. I'd probably be accused of hauling contraband, and my vehicle would be confiscated.

As soon as I get home with found materials, I inspect them for sharp edges, which I mash down right away. Then I soak the materials in hot soapy water, rinse them, and dry them on outdoor racks. If I do jab myself, I don't want the jabber to be dirty.

SPAM CAN TANK

One day, while I was browsing in the tin can bin at our recycling center, I found a Spam can with both ends removed. It was a little dented, but I retrieved it and was just about to square it up when I noticed that it looked exactly like a tank. My mind flashed back to myriad connections between Spam and the army, and I immediately started a turret search. A small rubber-cement can caught my eye, and soon I was tinkering away on my Patton-ed tank toy. Make your own by using the tips that follow.

A rubber-cement can is the perfect shape and diameter for a turret, but it's too tall. To cut it down to size, first scratch a line around the middle of the can. Next, scratch lines to indicate three tabs for fastening the turret to the tank.

Cutting a can through its sidewall without damaging the ends isn't easy. You should make a rough cut first, and then trim the cut down to the scratched guide line. The tool to use first is a knife. Luckily for me, the best knifemaker in the world,

Rob Hudson, made me a "Maryland Can Opener." (The name is a joke—a spoof on a long, deadly stiletto called an Arkansas Toothpick. My version is deadly only on cans.)

Place the can on its side and brace it against the back of your work surface. Slowly and very carefully, push the point of the knife straight in, about 1/2" (1.3 cm) to the waste side of the guide line. Turn the can just a bit and slice it, gently pressing in and downward. Repeat, slicing away from yourself, 1/2" or so at a time, until only 1" (2.5 cm) of uncut sidewall remains.

At this point, the can will be pretty flimsy, so bend it open and finish the cut with shears. Shears won't work for the initial slicing because the can's sidewall is too stiff to bend away from the thick blade.

It will be much easier to use the shears or snips now, but don't try to cut right up to your marked guide line yet. It's best to cut 1/8" (3 mm) or so away from the guide line first and then cut along the line itself on the next pass. Why? Because as you cut, the waste metal curls up and gets in the way of the shears, making it hard to cut accurately. The thinner that metal strip is, the easier it rolls up.

When you're finished with the second pass, use the shears to trim the

metal to the cutting line. (You'll probably find it easier to use both right- and left-hand cutting shears to cut the circle and tabs accurately.)

Hold the cut turret in position and scratch a line on the Spam can body where you want to insert the first turret tab. Then use a knife to carefully cut the marked slot in the body. Insert the point of the blade at one end of the line and push straight in with the cutting edge facing the center of the scratched guide line. When the blade is about 1/4" (6 mm) deep, remove it, reposition the point at the other end of the guide line, and push the blade in again until the cut is complete. By using this technique, you'll end up with a tab slot that has square ends. Cutting the slot with a single slice of the blade would leave you with a slot shaped like the blade's triangular cross section.

Insert the first tab, mark the second slot, and cut that slot. (You can mark and cut the three slots all at once, but doing them one at a time is safer.) Repeat to mark the third slot. Insert the three tabs, press the

turret down snugly, and bend the tabs over inside the tank body.

Stand your tank on its side on a scrap of 1/2"-thick (1.3 cm) board and trace around the tank's inside with a pencil. Flip the can over and trace the other opening onto another scrap of wood. Cut out these wooden shapes and tack some bottle caps on them to make roller gears. A few upholstery tacks complete the illusion.

Insert the wooden side pieces just inside the rims of the Spam can and hold them in place by hammering small brads through the metal.

I was grubbing around in one of my boxes of odd parts looking for a suitable gun, when I chanced upon an old ball-point pen. Inspiration struck. The pen is mightier than the sword, or so I've always heard; surely it's mightier than a cannon. I drilled a hole in the turret to accommodate a working ball-point and—voila!—an ideal desk accessory for the Pen-tagon or a think tank.

■ *Camouflage Tank,* Bobby Hansson
This toy tank was made from three carefully opened evaporated-milk cans (see below). To disguise the fact that this model required a lot of solder and to add a different look, I painted it with "Desert Storm Camouflage Paint." After mastering the Spam Can Tank, you should be able to make this version just by using the pictures as guides.

■ *Folgers Shoulder Soldier Holder*
Bobby Hansson
If someone hands out free posters of Napoleon, and you don't want to fold yours, roll it up and put it in this handy carrier. The stack of soldered coffee cans will keep rolled-up maps or posters safe and dry. The end of a juice can, fastened down with lunch-box hardware, seals the end, and a guitar strap makes a handy shoulder strap.

■ *Adam & Eve,* Dick Crenson
Dick made this delightfully wacky piece for an art show at the Pleasant Valley Recycling Center. Eve is 25" (63.5 cm) tall. Dick told me with a chuckle that he took this photo because he was afraid someone might recycle the piece after the show was over. I'm happy to report that no one did.

■ *Lamp,* Artist unknown
My friend Betty Oliver was given this lamp from the Dominican Republic.

■ *Can-estoga Wagon,* Bobby Hansson
To make this wagon, I first removed the top and bottom of a coffee can. Then I placed a piece of 2 x 4 inside and used a wooden mallet to bash one side of the can flat. I made the seat by cutting and bending the detached lid at a right angle and soldering it in place. For the axle and wheels, I used a rolled-up tin tube and four jar lids, inserting the axles through holes poked in the wagon body. The ox was cut from a flat piece of tin can and bent so it would stand. Most of the paint on this model was sanded off because the printing was deemed inappropriate.

■ *Lamp,* Artist unknown
This crusty old lamp belongs to Dan Van Allen, who found it in Haiti. It works just like the Dominican lamp shown above it, but the tin can fuel reservoir seems less scary than a light bulb filled with kerosene.

■ *Lamp* (right), Bobby Hansson

Here's my adaptation of the Dominican lamp shown on the opposite page. I started by drilling a hole in the base of the light bulb and removing the black insulating material. Next, I opened a spray can and cut it to 1" (2.5 cm) in height, leaving three narrow strips of material which continued to the top of the can. I placed the bulb in the can bottom and wrapped the strips to hold the bulb in place. The wick is held by a metal tube inserted into a hole in a bottle cap.

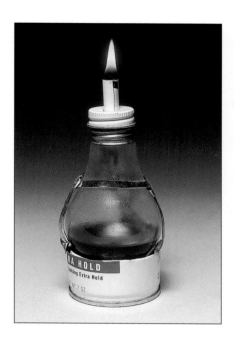

■ *Lamp* (far right)

Bobby Hansson

A fancier version of the lamp shown next to it, this project includes an aerosol-spray can as a base. I cut the can to about 4" (10.2 cm) in height and rolled the top edge around a piece of baling wire. The wire holds the bulb in place and secures the reflector, which I made from the bottom of an aerosol-spray can. A decorative strip from a coffee can is wrapped around the bottom of the spray-paint can.

■ *Camera*, Bobby Hansson

This Kodak commemorative can, which was made to introduce a new type of film, looked to me like a camera body. I soldered a small coffee-sample can to the bottom to serve as a lens. Note the small pinhole in the center of this can. To make a functioning pinhole camera, paint the inside of the coffee can matte black. Step into a darkroom, take the lid off the rectangular can, fasten a sheet of film inside (emulsion side facing the lens), and cover the hole until you're ready to shoot. Uncover the hole for the exposure. The hole made by an ordinary push pin is roughly equal to an f/90 lens opening.

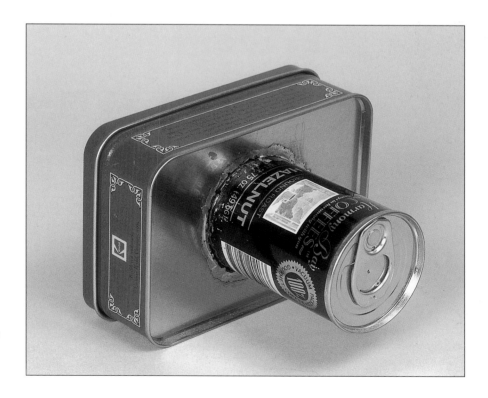

■ *Lamp*
Bobby Hansson
This tin truck was turned into a lamp by adding a homemade tin shade. A folded strip of tin holds the shade in place on the lamp harp.

■ *Coffee Can Lamp*, Christopher Anna
This space-age desk lamp lets you redecorate your desk space as easily as you switch brands of java.

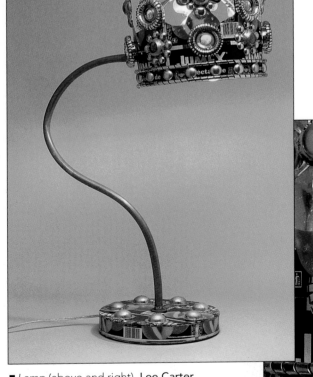

■ *Lamp* (above and right), Lee Carter
Here's a heavily encrusted lamp from the Mexican workshop of Lee Carter.

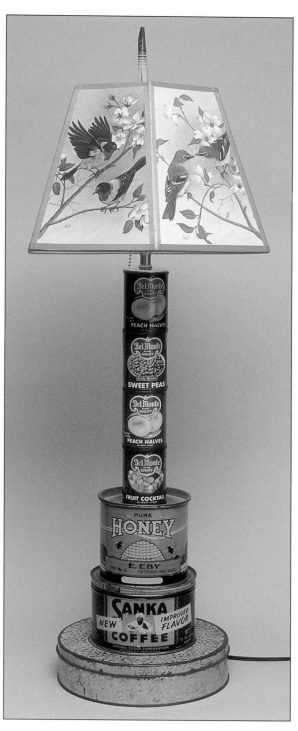

■ *The Tin-Fanny Lamp* (above)
Ricki Boscarino
The base of this piece, which supports
a fabulous, fanciful, and flamboyant
lamp, was made from a Texaco oil can
filled with cement. Notice the ceramic
tiles in the cement.

■ *Lamp* (right)
Roberta and David Williamson
This towering tin can lamp was a
collaborative effort.

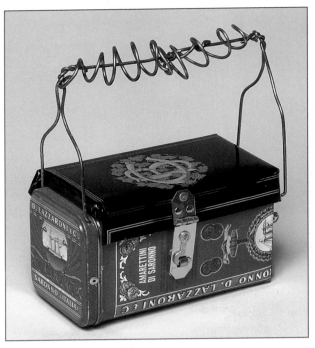

■ *Ladies Handbag*, **Bobby Hansson**
This ladies' handbag is used for carrying opera glasses
to the theater.

■ *Briefcases* (left and above)
Bobby Hansson
Soon after I started doing research for
this book, I found that I needed something in which to keep my notes and
papers, so I decided to make myself a
tin can briefcase. The finished project
was so much fun to use that I made two
more (one is shown on page 33), each
slightly different. Because the round
cans I wanted to use for the tops never
seemed to be as long as the square
one-gallon (3.8 l) cans I used for the
bodies, I had to make the tops by riveting together two or even three overlapping cans. I salvaged the hinges and
hasps from rusted-out tackle boxes for
two of the boxes and used new hardware on the third.

■ *Grand Opening Purse for Zette*
Bobby Hansson
Owned by Zette Emmons, who
sometimes organizes film festivals,
this evening bag includes 16mm film
cans as end pieces for the lid.

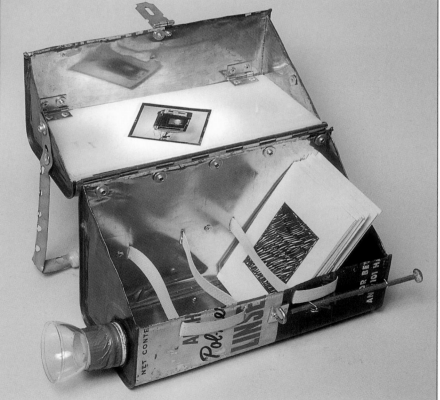

■ *Portable Light Box*
Ada Tolla Le Ray and Giuseppe Lignano
The two Italian architects who made
this light box and who own Lo/Tec in
New York City, design and build marvelous objects using recycled material,
much of which they pick up on the
street. After they saw my lunch-box-
shaped briefcase, they made this
portable light box for viewing photo-
graphic transparencies. Like everything
they create, the box looks strange but
works very well. When it's open, the
handle serves as a leg to hold the white
plastic light source level. The plastic
strapping dividers make compartments
for storing photos in an orderly fashion.
The spout hole holds a magnifying
loupe. When you need new batteries,
it's easy figure out where to put them.

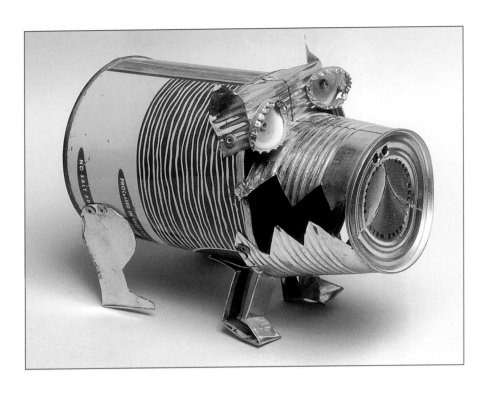

■ *Untitled,* Bobby Hansson
The head of this sad-eyed fellow was made from the same type of milk cans as the tank on page 37. A little deft cutting and bending, an ice-picked nose, and a mouth dented in with a butter knife complement the bottle-cap eyes.

■ *The Medusa,* Dick Crenson
The artist made this piece (which is not a can of worms) from a tomato-sauce can and several cat-food can covers.

Chapter Four

Working with Deconstructed Cans

■ *Figures,* **Janet Cooper**

LEARNING TO cut sheet metal is like learning calligraphy. Anyone can draw a recognizable letter on the first try, but copying the Declaration of Independence takes a bit of practice. You'll need to spend some time getting used to different tin snips, shears, and knives.

The great thing about tin can art is that the materials are free and easy to find. If you recycle, tin cans are worth just as much after they're hacked up as they are intact, so practice your cuts before you turn in your empty cans. To be nice, put all your sharp scraps in a large can and fold it shut before tossing it in the recycling bin.

Practice freehand cuts and practice following scribed lines on metal. Practice on flat tin, curved tin, and cans with corrugated or ribbed sections. Stay calm, go slow, be patient, and be careful. Start by practicing on a flat sheet; can lids are a good material for these cuts. If you're not handy with scissors, you'll find it helpful to practice cutting stiff paper or index cards first. You'll be making cardboard patterns later anyway, so why not get the hang of cutting cardboard now?

TIN-SNIP TIPS

—Keep your snips moving forward. If you back up, you'll create a burr on the metal.

—Control the piece with your tin-holding hand, turning the material into the cutting edge of the blades.

—Stop often to pick up your scraps. You'll be making a safer, neater working environment, and taking breaks will help prevent scissor-hand cramps and carpal-tunnel weirdness.

—When you're cutting curved cans, cut 1/2" (1.3 cm) or more wide of the mark first. Then cut 1/4" to 1/8" (6 mm to 3 mm) wide, and finally, try for the line. The reason for making three cuts instead of one will become obvious the first time you try to cut an intact can with snips. When you cut metal with these tools, the waste metal will rise up and curl. On a flat piece this isn't a problem, but when you're cutting a curved piece, the curled metal will jam against the back of the blade and handle. To prevent this, hack off the waste first to get a rough outline. Refine the cut next, and then cut to the finished

shape. This may sound three times as difficult, but it's really ten times as easy.

—Many smart people prevent serious cuts by wearing gloves while working with sharp metal. On the other hand, the smartest man I know, Peter Ross, is a blacksmith who never wears gloves. His theory is that a glove won't stop a 2000°F (1093°C) hunk of iron from burning you, and if you start to think that it will, you might do something careless—and dangerous. Peter prefers to be aware of his vulnerability and to use his brain to prevent mishaps. I advise you to do everything possible to be careful. If you want to use gloves, please do, but be aware that if you're careless, sharp tin will slice through them.

—You don't have to suffer to make art. If you notice your knuckles getting white, stop. Breath deeply. Don't ever force the tool; let it do the work. I always listen to music while I'm working. When I start to get tense, I play a tape of Willie Nelson singing Hoagy Carmichael tunes, and I'm a bliss-ninny in no time. It's great to make a wonderful object, but if you don't enjoy the process, you'll spend too much of your life being stressed out and grumpy. You can have fun and still be serious about your work.

■ *Sheik,* Sue Eyet
A sheik on horseback rides roughshod across a winged heart.

TIN WHISTLE

John Daniels, a blacksmith from Stafford, Virginia, showed me how to make the whistles pictured below. John sometimes uses the steel strapping material with which truckers secure cargo, but we'll make our whistle out of tin can material, using snips, a dowel, and a small pair of pliers.

Start by snipping a 5/8"-wide (1.6 cm) strip of tin. From it, cut one 3-1/4"-long (8.2 cm) piece and one 1-1/4"-long (3.2 cm) piece. Because you'll be putting the whistle in your mouth, you'll need to remove any burrs or sharp edges right away.

To start shaping the curved section of the whistle, make a 5° bend 1/8" (3 mm) from the end of the longer piece. Continue curving the strip by wrapping it tightly around a 5/8"-diameter (1.6 cm) dowel. Next, measure 1-1/8" (2.8 cm) from the unbent end and bend up sharply, as shown in the photo above.

Place the flat portion of the bent piece across the short strip, centering it carefully. Bend the ends of the short strip up and fold them over the piece on top. Then fold the protruding end of the long piece up and over the bent ends of the short piece.

■ *Hog Clock,* Bobby Hansson
A city slicker visiting the country saw a farmer and a pig standing under an apple tree. The farmer was lifting the swine up over his head so that the pig could reach an apple. When the pig bit off the fruit, the farmer lowered him to the ground and watched him eat it. The farmer then hoisted the pig again, let the beast harvest another apple, and lowered him once more. After watching this process being enacted a dozen times, the visitor couldn't refrain from suggesting that a lot of time could be saved by shaking the apples out of the tree and letting the hog browse among them. The farmer gazed with amusement at the city boy and said, "Time doesn't mean anything to a pig."

   Thus it is that this ham-can chronometer is equipped with a pig-tail, but has no numbers.

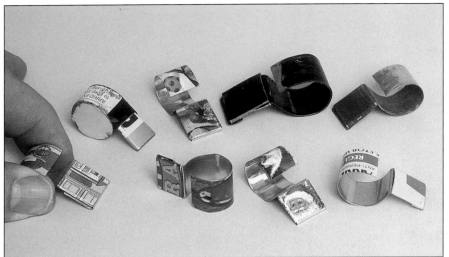

The air passage created by the short piece should line up with the end of the round loop to give you the equivalent of a policeman's whistle without sides. When you hold the whistle as shown in the photo above, your thumb and forefinger will act as its sides. Notice the yellow whistle in this photo; it includes optional side pieces, for no-hands tweeting! With practice, you'll be able to eyeball the length of the strips and make all the bends with your fingers.

When its sections are lined up correctly, this whistle can be heard 100 yards (91.4 m) away. I've found that its sound goes a long way toward convincing a classroom full of rowdy boys that learning about art can be fun.

Make minor adjustments until you get earsplitting results. Once you get the hang of making these whistles, try different widths and lengths of tin and look for cans with printing that will make amusing designs.

■ *Tin Bells,* Wallace M. Yater

Wally made these little bells as Christmas decorations when he was 13 years old. A lot of careful planning was required to get their shapes to nest together as nicely as they do, and the neat handiwork gives some indication that this lad would grow to be a remarkable craftsman.

SOLDERING

Soldering tin cans is easy because the tin coating on them is very compatible with solder. In fact, most tin cans are made with soldered seams and lids.

The best way to learn how to solder is to start by reading as much as you can on the topic; many books are available. Then talk to and watch people who do a lot of soldering. And practice! There are many ways to clean metal, many types of flux and solder, many ways to heat joints, and many ways to hold the joint together as the solder cools. With these facts in mind, I'll offer some hints that have been helpful to me.

away from your soldering bench and avoid slopping it around when you apply it.

Always take the time to clean the pieces to be joined. Scrape and sand away all paint, shellac, grease, and oil. A piece of cloth-backed abrasive paper can do this job quickly.

Cut and bend the pieces to fit together snugly. Then apply flux to the areas where you want the solder to flow, as shown in the photo at the top of the page. This acidic substance helps clean the metal and helps solder flow evenly. It will also corrode your tools and stain the metal in your project, so keep flux

If possible, clamp the pieces together before soldering. The tighter the fit, the stronger the solder joint will be. Although solder will fill gaps, the bond between pieces won't be as strong if these gaps exist.

Using your soldering iron, heat the joint to the melting point of the solder and allow the solder to flow over and around it. When you're soldering long seams, tack the pieces together with a little solder

■ *Portrait of a Saint,* **Dick Crenson**

■ *Bracelet,* **Bobby Hansson**
Made from a pair of beer cans wired to a soda-can wristband, this bracelet has sharp edges and is more fun to look at than to wear.

here and there before running solder all along the seam. Be sure not to disturb the joint until the solder has cooled and hardened, or the joint will be weak and may even come apart.

Unless you've used an acid-free flux, which will wash off with soapy water, you'll need to neutralize the acid in the flux by washing the finished solder job thoroughly with a solution of baking soda and water. If you don't, the piece may rust later on. Water-rinsing or just wiping the metal will only dilute the acid and slow down its corrosive action.

RIVETING

Pop rivet tools, available at your local hardware store, come with complete instructions. They're generally used with rivets that are 1/8" (3 mm) in diameter and with a 1/8" grip range. To use the tool, hold the work together securely and drill a 1/8" hole through the pieces to be fastened. Insert the nail-like end of the rivet as far as it will go into the nosepiece of the tool and place the other end of the rivet into the hole. Squeeze the handle of the gun firmly until the nail-like end snaps off. To flatten the rivet even more, just place your work over any heavy iron object and strike the top of the rivet with a hammer.

Split rivets are inserted in a different manner. Place the rivet in the drilled hole and split the legs apart with a screwdriver. Fold the legs down and strike them with a hammer until they grip the metal securely.

ORGANIZING

Stan VanDerBeek was a marvelous fellow who made animated underground movies using cutouts that he

■ *Sign and Pocket Watch* **Sue Eyet** The 27"-high (68.6 cm) pocket watch acts as a Sign of the Times—for a Clock Shoppe. Sue nailed cut-out sheets of tin onto a 3/4"-thick (1.9 cm) plywood backing and made the numbers from sheet copper. As you can see in the close-up below, Sue makes great hands.

■ *Masks,* Collection of W.H. Bailey and Laurie B. Eichengreen
That an artist could have produced these masks—found in Bolivia—from cut-up tin cans is incredible, but as you can see in the view of the back of the horned mask, the can labels are still visible. Getting tin to stretch into these smooth-domed shapes isn't easy. It must either be done instantaneously, with a huge amount of force, in the same way that car fenders are stamped out, or very slowly and gradually with a lot of careful, gentle taps from a mallet, to avoid creases and wrinkles.

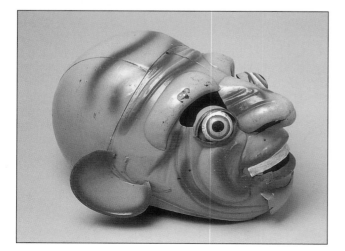

manipulated in fantastic ways. He built his house and movie-drome theater out of salvaged, found, and recycled materials, and he loved to cut up tin cans and juxtapose images to create amusing objects. Stan taught me that unless you keep your materials organized and accessible, they can hurt you—instead of enriching your palette—by cluttering up your work space. Stan had boxes full of images that he'd sorted into categories.

When I'm waiting for my soldering iron to heat up or for a painted item to dry, or, more often, for my muse to summon me, I like to sort through cans, cut them up, and file the cut pieces in my tin bins. I have several small boxes filled with tin words like "guaranteed," "caution," "pure," "improved," and "warning."

I keep pieces with animal pictures in one bin and food pictures in another. Plain colored scraps are divided into warm and cool colors. When I have a project in mind, I cover the table with a large sheet of paper and turn out a bin's contents onto it. When I've found just the right piece, I pick the paper up by two of its edges and dump everything back into the bin. Try to use something other than your bare hands to do your rummaging; in the excitement of discovery, it's easy to cut yourself.

Having things organized also helps when you're foraging. If you know that you already have 73 tuna cans, you'll probably want to leave one for someone else to glean. It goes without saying, however, that you can't get enough of some cans. Kikkoman soy-sauce cans are a great example. I love that beehive texture and those great colors.

HARRIETTE ESTEL BERMAN

My recent work is a series entitled "A Pedestal for a Woman to Stand On." The minimal cube shape is a self-imposed restriction presenting a formal sculptural approach. The use of squares also represents my home and the fundamental structure of quilt patterns. I used toy tin houses as raw materials, disassembling, folding, fabricating, and riveting them together. These pieces of a "home" are an effort to reconstruct my life. They symbolize both autobiographical content and the complexity of women's lives in today's society.

The title offers an important insight into the meaning of the work; it refers to the quilt pattern used in the piece and makes a visual and conceptual connection to the history of quilts—historically, one of the few outlets for creative women.

Hourglass Variation (shown above) evolves from issues that are both highly personal and yet of concern to all women in our society. The edges were cut with pinking shears (a reference to fabric) and turned up so that even the thought of standing on this scale would be extra painful! The quilt pattern is an "hourglass variation," a visual pun on the stereotype of a woman's figure. The words on the dial ("sugar," "tomato," and "cupcake," for example) are words used to describe both food and women.

■ *Hourglass Variation: The Scale of Torture,* Harriette Estel Berman
Harriette is a serious artist with a sense of humor.

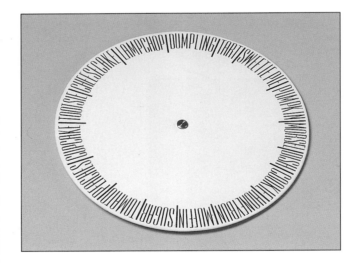

The dial of the scale (shown to the left) includes words commonly used to describe women as well as food. The top surface of the scale (below) shows painful edges that were cut with pinking shears. Note the attention to detail. The rivets in red areas, for example, are painted to match the background.

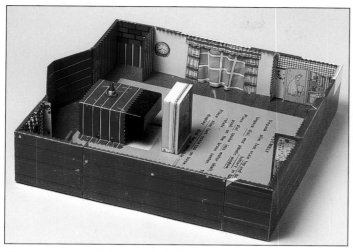

The base was made from a tin doll house. Note the assembly instructions on the floor and the dial motor on the left-hand side.

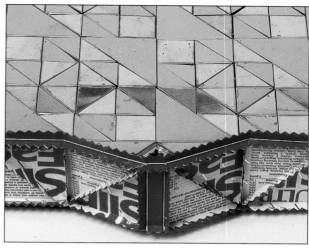

On the underside of the top section, the tin is folded in what quilters call an "hourglass" pattern.

■ *Folded Tin Can Necklace*
Natalie Frost Weeks
The artist made this folded tin can necklace by using a dappling block to shape the gold pieces. The gold color is the original color of the inside of the can. Natalie's bold, honest treatment of the material results in a very attractive piece.

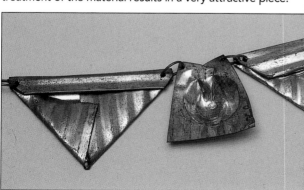

■ *Binoculars,* Ruth Abraham Fisher
The artist invented and built this humorous and slightly naughty pair of bird-watching binoculars. Ruth lives in London, where young women are sometimes called "birds." The can from The Virgin Trading Company features a pinup girl, and the lenses are made to resemble bulging eyes. All the brass wire in the handle suggests that only a viewer with a lot of brass would use these spyglasses. The wire is also somewhat twisted.

■ *Figures,* Janet Cooper
The artist who created this nostalgic little tribe (and the figures shown on page 45) was a potter for many years and valued making objects from raw earth. She now makes personal adornments (mostly jewelry) from society's throw-aways, notably vintage bottle caps. The figures' movable arms and legs were snipped from vintage tin cans.

■ Left: *Dr. Blue,* Right: *Commander Red Fosters,* Sean Duffy
The artist stuffed and laced these animals together with yarn. They're made from cut-up aluminum cans, not tin, but are so adorable that I don't care if they can't be picked up with a magnet. The printed sections are positioned judiciously to create a bold overall look.

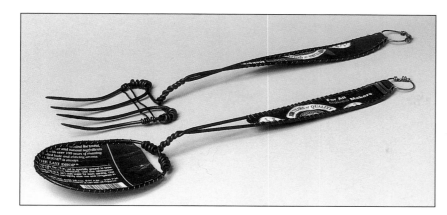

ELLEN WIESKE

The question of what is "precious" or important as a material has drawn my attention away from gold, silver, and diamonds to steel wire, tin, and copper. Common, plentiful, and usually overlooked materials such as these, handled with artistic temperament and integrity, have raised these questions for me: What will this do to the pieces? To their maker? To the viewer?

■ *Tableware,* Ellen Wieske
Ellen, who has a Masters degree in metalsmithing, has introduced an element of humor to this tableware by incorporating the words printed on the tin can sections.

■ *Footstool with Tin Strips,* Sue Eyet
The body of this little footstool, bought at an auction, was in terrible shape and was missing a seat. Sue refinished the wood and wove a twill-pattern tin seat. The narrow strips were cut from lighter-fluid cans and then hooked and soldered together. Those running in one direction are positioned with their printed sides up; those running in the other direction are face down. The stool is very strong and quite springy and comfortable.

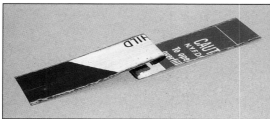

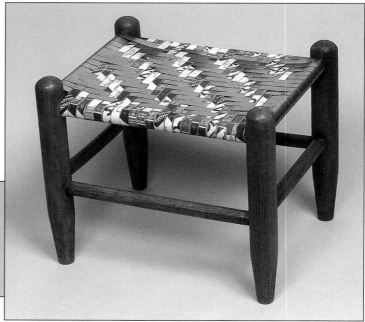

RANDALL CLEAVER

My work combines discarded items to create time-keeping artifacts. Their utility and motion involve the viewer in their complexity of forms, textures, relationships, and humor.

Creating with found objects started as an inexpensive way to obtain materials, but the objects themselves soon became a source of inspiration. I try to give my viewer the sense that the parts were manufactured specifically to form the object, so the various parts transcend what they once were.

As a body of work, the clocks represent a conglomeration of ideas and concepts I've had over the years: actual, as opposed to implied motion; machine sounds emanating from the pieces; humor; functionality; the sense of history in timepieces; and the near obsession our society has with time. The clocks also provide an archetypal starting point from which the viewer can move more deeply into my works.

 *Reliquary of Time Past*, Randall Cleaver

The artist's powerful, unique style makes *Reliquary of Time Past* a remarkable work. Although he purchased the clockworks, Cleaver made everything else, including the hinges and hasps.

Randall's working style is intuitive. He starts with a germ of an idea or a particular found object that suggests a piece, but as the piece grows, different relationships emerge and the form of the work responds to these emerging discoveries.

CHAPTER FIVE:

ARTISTS AT WORK

■ *Heavy Ty*
Harvey Crabclaw
The image of a giant ear of corn on a popcorn can inspired the artist to make this frame for a Ty Cobb baseball-card photo. Because Ty was nicknamed "The Georgia Peach," Harvey included peach can parts. The "Ty" on the frame was taken from a can with the printed words "Heavy Duty."

While he worked, Harvey tested me with corny riddles. What did Ty do in his high school marching band? Played the cornet by ear. What does Ty do when he strikes out? Grits his teeth and says "Shucks!" What did Ty do in the army? He was a Colonel and stalked the enemy.

I'M FASCINATED by the expertise with which artists handle their tools, and I love to gain insights into the ways in which their minds operate. Not all artists are willing to be observed at work, so for those readers who don't have easy access to gregarious artists, here are some glimpses of a few doing what they do.

HARVEY CRABCLAW

Harvey Crabclaw is an itinerant folk artist whose studio is an exhausted school bus, which he tows across the country behind an ancient Peterbilt semi-truck tractor. Harvey spends most of his time in New Mexico; his picture frames show the influence of Mexican tinware, cowboy leatherwork, and American Indian patterns.

This artist's work habits are as unique as his lifestyle. He'll spend a couple of months in New Mexico, working at one of his myriad vocations (including tattooing, sign painting, T-shirt printing, and woodcarving), and when he starts getting restless, he hitches up his rig, which he calls the "Busted Bus Tinworks," and drives. Last year he came to visit me in Maryland, at the head of the Chesapeake Bay, and parked in the backyard for a couple of weeks.

Harvey loves to spend time looking around or, as he puts it, "soaking up notions—bits and pieces of things that make me smile." My wife Maggie and I have hundreds of books, and Harvey spent hours browsing through them, searching for pictures for his frames. He wanted to hear the stories about the people who caught his eye. He

marked a couple of dozen pictures he needed, and we photographed color-print copies of them out in the yard. Twenty-four hours later, we picked up the 4" (10.2 cm) prints from the supermarket.

Harvey's good at talking while he works, a skill he learned as a tattooist to distract his customers from any discomfort they might experience. Harvey knew I was writing about tinwork (that was his excuse to visit) and was eager to share his expertise.

"Let's see if I can conjure up something cute for your readers."

FRAME

One of the photos we copied was a detail of a self-portrait by an artist named Joe Coleman, who according to Harvey, did a performance in Boston by taping fireworks to his chest and setting them off. A minor panic resulted at this performance, and Coleman received a summons for "burning a dwelling house." The day after we copied Coleman's portrait, we found a one-gallon (3.8 l) can that had "Coleman Lantern Fuel" printed on its side. Harvey was exultant.

Coleman had written a book entitled *Cosmic Retribution*. "This,"

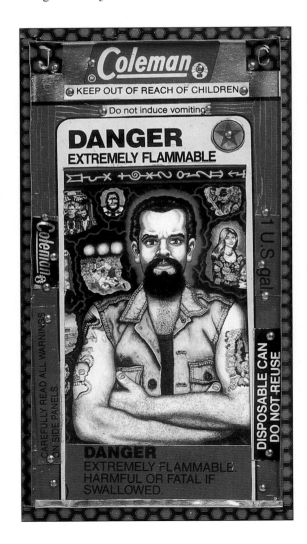

■ *Framed Photo of Joe Coleman*
Harvey Crabclaw

■ *Mrs. Nez,* Harvey Crabclaw
Decorated concentric circles seem appropriate for framing a photo of this potter, who lives near Santa Fe. The scalloped edges were made with the crimping tool shown on pages 14 and 15.

■ *Rosie Can,* Harvey Crabclaw
This is a tribute to all the women who pitched in to help win the war. Harvey fastened together the bottom of a soy-sauce can and the top of a coffee can with—what else?—a rivet.

claimed Harvey," is Karmic Contribution! Look what it says on the can—'extremely flammable,' and Coleman's name is printed all over it!"

I washed the can inside and out with hot soapy water, and Harvey took off the top and bottom with a can opener. "Check this out," he said, as he used the can opener sideways, with the wheel cutting through the wall of the can rather than the top. "If you're going to use the can flattened out, this saves cutting through the rim, and you can use the bottom with the rim around it for a frame, too."

Next, Harvey slit the can from top to bottom, along the sidewall seam. "Unless you plan to include the seam for some reason, it's the best place to split," he ruled.

Flattening the can gently on the table, he picked up the photo and tried out several positions. The can's side panel, which included the word "Coleman" and the "flammable" warning, was the obvious choice. Harvey cut out the top and both sides of the frame in one piece. The bottom piece had to be added because the photo was too large. For this extra piece, Harvey chose a section of the container that reiterated the "flammable" warning.

"I see this guy as flammable in more ways than one," he intoned gravely, and immediately started to chuckle as he snipped out a section of the instructions reading "Do not induce vomiting."

"Coleman also got in trouble with the A.S.P.C.A. for swallowing live mice," he noted, "so this is appropriate."

After he'd snipped all the elements he wanted from the can, Harvey arranged them on top of the photo until he was happy with the look. Almost. "What's lacking here," he announced, "is a Kikkoman border," referring, of course, to the beehive pattern on Kikkoman soysauce cans.

We found a piece of wood about 3/8" (1 cm) thick and cut it slightly larger than the frame. Harvey bent the Kikkoman strips into U-shaped channels, fitted them around the edges of the board, and nailed them in place with brass brads—a "look" borrowed from John Grant (see page 11). Next, he glued the photo down with airplane glue and nailed the flat pieces in place with 1/2" (1.3 cm) brads, bending the brads over where they protruded from the wood backing.

To facilitate nailing, Harvey used an ice pick to punch tiny holes in the tin. To his dismay, one of the brads somehow followed the grain of the wood backing while Harvey was nailing down one of the tiny warning strips, shifting the position of the tin piece as it did. Because the element was so small, he decided to accept the accidental effect, but declared, "We must eliminate the possibility of a reoccurrence of this phenomena."

From that point on, he taped the tiny strips firmly in place with transparent packing tape before making the holes and nailing the strips down; this worked well. Harvey's instant invention of a technique to solve this problem is one example of why I admire people like him and Jose de Creeft. They manage to combine a relaxed mood and real enjoyment of their work with constant vigilance, pursuit of excellence, and flexibility.

The initials "J" and "C" in the uppermost corners of the frame were stamped from the back with reverse letter punches (see page 94 for more information on punches), and these raised areas were then sanded down to the bare metal.

This frame works well for several reasons; the red color in the painting even matches the tin can. When

I told Harvey that I once met Joe Coleman briefly, and that we'd both appeared in a film (*Shadows in the City*, by Ari Roussimoff), he presented the frame to me with a mock-formal flourish.

Harvey fits the frames he makes to the pictures they surround. He does this in different ways, depending on the subject and his mood. He considers both verbal and visual elements; as he sees it, homage or sacrilege are both valid artistic approaches.

■ *Mashed Mask of Agamemnon*
Harvey Crabclaw

■ *Big Star,* Harvey Crabclaw
When Harvey found this photo I'd taken, he couldn't resist making this frame.

■ *Peary Aprés Pole Jaunt*
Harvey Crabclaw
This photo was taken right after Robert Peary said he'd discovered the North Pole. Several Arctic explorers died when canned goods contaminated with lead poisoned them. The lead pencil in this piece, which looks like a pole on the tin can "globe," asks "2B?" The rim of bottle-cap dogs is a tribute to the noble beasts who hauled sleds across those frozen miles.

HARVEY CRABCLAW

Working with old tin cans isn't like working with store-bought stuff. It's similar to making a quilt from grandpa's necktie, grandma's apron, and dad's old work shirt instead of from new materials. Used and rediscovered objects have a history—a patina. They seem to be more poignant than brand-new objects. What I find when I look at them really turns me on. Using found objects just because they're free, or just to be a recycler, eliminates part of the excitement and would be like eating grits without pepper.

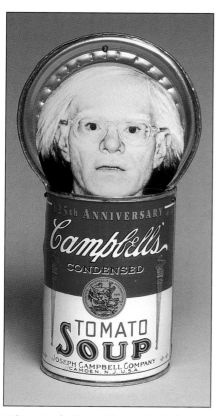

■ *Saint Andy Rising,* **Harvey Crabclaw** Haloed by the lid, Warhol emerges triumphant from his soup can.

■ *Icon Tina Turner,* **Harvey Crabclaw** The title refers to Tina's early career and honors the way in which she has transcended it to become the powerful symbol she is today.

RICHARD HADDICK

Richard Haddick worked in the sheet-metal trade for many years before he became a full-time tinsmith. He can make virtually any shape in tin and is especially talented at drawing and at creating patterns. When a museum gift shop wants to order 50 candlesticks, for example, Richard makes a tin template for each section of the job. He holds these templates down on a sheet of tin, scratches the outline with an awl, and then cuts the sections out. Careful planning helps him get as many parts as possible out of one sheet of tin.

As well as shears, pliers, and hammers, Richard has several big bending and cutting machines. They're all hand- or foot-powered, and some are more than one hundred years old. Although he used electric and hydraulic machines when he made sheet-metal roofs and ducts, he uses traditional hand methods for all his tinwork. In fact, his only modern tool is an electric soldering iron.

Richard usually works with flat sheets of new tin, but he kindly agreed to demonstrate how many designs could be made from tin cans. I went to his studio with a bagful of cans and a sheet of printed tin I'd picked up at a can factory before it was rolled up to make cans. It was fun watching Richard work. He's a very centered person and a meticulous craftsman.

■ *Candle Extinguishers* (model and project), **Richard Haddick**

CANDLE EXTINGUISHER

A small cone with a looped handle was near Richard's work table, and I asked him to show me how he'd made it.

Richard chose an olive-oil can from my bag and used large shears to cut off the top and bottom. Then he cut down along the seam, opened the can, and flattened it. Next, he set his pattern on the can, centering it carefully over the printed medallion. "Placement is the key," he explained. For the loop handle, he chose a lettered section.

Next, he scratched an outline around the pattern and cut it out. Then he used a bar folder to fold over the edges of the handle section. The handle's double-thick seam stiffens it and creates a smooth, safe edge. "I could also fold the edge by clamping it between two blocks of wood, but this is a lot faster," he explained.

Using a rawhide mallet, he shaped the cone by hammering the tin over a pointed mandrill, working toward the middle, first from one edge and then from the other. Richard used a wooden mallet and an iron bar to start a bend at the end of the handle loop, and he finished the bend with his fingers.

Using a green scrubbing pad and holding a ruler along the edge to get a neat straight line, he scraped the paint off the areas that he was going to solder. Then he soldered the seam of the cone. After he carefully lined up the handle with the seam and attached it, I photographed the little candle extinguisher next to the model. We liked the way the logo design looked and decided it didn't need the indented lines included on the model.

DUSTPANS

Richard also showed me how he makes his mini-dustpans. He started by cutting the top and bottom off an old gasoline can and flattening the can out. He positioned his pattern so the can's eagle design would be centered, traced around the pattern with a sharp scriber, and cut the tin out with shears. His pattern

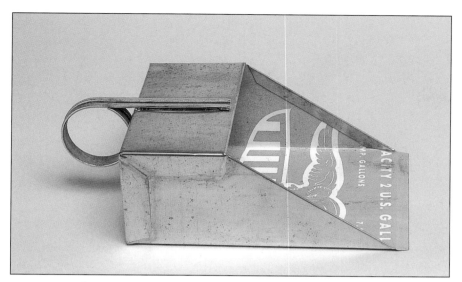

■ *Dustpan,* **Richard Haddick**

■ *Dustpan Pattern,* **Richard Haddick**

is designed so that one piece of tin folds up to make the dustpan, complete with tabs to reinforce the corner joints.

Richard removed the paint from the areas to be soldered and applied liquid flux to them. As he soldered, he used his long, pointed scriber to hold the seams together tightly, explaining that the tighter the pieces were held together, the stronger the solder joint would be. (Richard sometimes uses rivets instead of solder to give the piece a different look.)

To make the handle, he cut a long thin strip of metal and folded the edges over twice to make a double seam for added strength and a nice smooth edge. To bend the strip, Richard used his bare hands to wrap the metal around a large dowel. He used a plastic mallet to hammer down the little wrinkles that always occur when a doubled-over sheet of tin is bent into a circle. (The metal was left on the dowel while he did this.) Then he soldered the handle onto the pan.

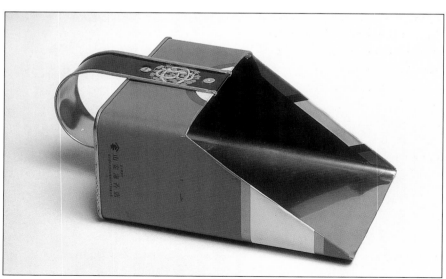

■ *Lazy Man's Dustpan,* **Bobby Hansson**
This version was made by cutting the top off a quart-sized square can and adding a handle.

■ *Sconces*, Richard Haddick
The town where tinsmith Richard Haddick lives isn't very big; he knows almost everyone. When word got around that he was using tin cans to make his lanterns and sconces, people started to bring all sorts of decorative cans to him. The olive-oil cans in these photos came from an Italian restaurant and from a pizza parlor. The animal pictures are from big dog-food containers given to Richard by the local animal shelter.

■ *Candleholder*, Richard Haddick

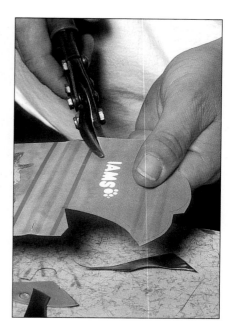

■ *Making a Sconce* (above and right)

To make a dog-food-can sconce, Richard starts by scribing the pattern. Then he blocks out the pattern by cutting wide of the scribed pattern line before he trims the metal to size. Before folding the metal, he uses a hammer to remove burrs on the cut edges.

■ On a table in Richard's work-shop are two of the huge dog-food cans given to him by the animal shelter. Some of the cut-out wall-sconce blanks are arranged on the shelf, and a lamp shade made from one of the cans hangs overhead.

PETER ROSS

Peter Ross is a blacksmith who is famous for making reproductions of Colonial tools, hardware, and cooking utensils. He lives with—and uses—the items he makes and often cooks in his fireplace, as was the custom in olden times.

TIN REFLECTOR OVEN

The tin reflector oven that Peter made while I visited with him is typical of those used from about 1780 through the middle of the nineteenth century. The oven is placed in front of the hearth, where the meat in it is cooked by radiation from the fire and by heat reflected from the shiny tin surface. To control the oven's distance from the fire, the cook just lifts the oven by the handles located on its top. These handles are soldered in place but are also riveted in case the heat of the fire softens the solder.

Take a look at the photos. You'll notice a wrought iron skewer in the oven's center; this can be fixed in a variety of positions by inserting the skewer handle into one of a series of small holes in the side of the oven.

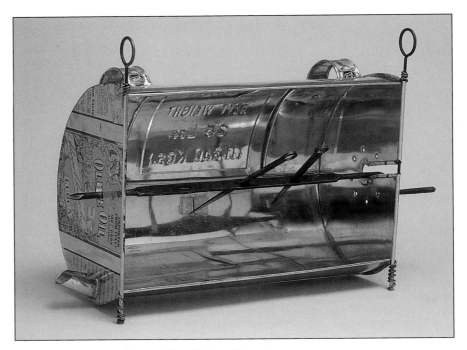

■ *Tin Reflector Oven* (front and back), **Peter Ross**

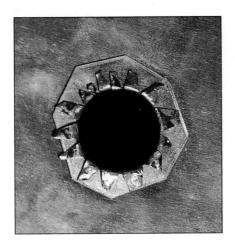

■ Detail of reinforced skewer hole

■ Bending the hinge flap around the wire on the door

■ Peter strikes the head of the rivet that reinforces the handle. The bottom of the rivet is supported by an iron bar.

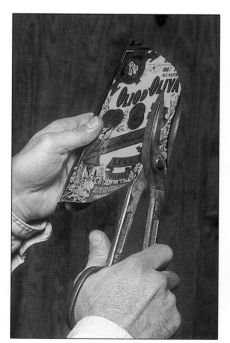

■ Notice how Peter holds the tin while he cuts it. Over-the-shoulder lighting makes it easy for him to see the scribed line.

■ Flattening the door wire over a wiring stake

The hinged door allows the cook to check the cooking meat and to baste it. A wire is encased all along the edge of the door to provide strength and to act as a pivot for the hinge.

Stiff wires, about as thick as a coat-hanger, are folded into the top and bottom edge of the oven opening; the protruding ends are wrapped around the leg wires, which are about twice as thick. The edges of the sides are rolled into tube shapes to hold the leg wires. When the leg wire on the slotted side of the oven is pulled up by its loop, the skewer—and spitted meat—can be removed.

Thinner wire, between 1/16" (1.5 mm) and 1/32" (.75 mm) in diameter, is used to reinforce and stiffen the trap door, the handles, and the rear legs.

The spout on the side is for pouring out the juices that collect in the bottom of the oven.

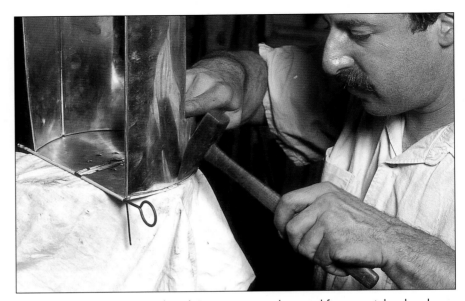

■ As he hammers the seam closed, Peter protects the metal from scratches by placing a piece of cloth on top of the anvil. Notice the protruding wire on the left; this will be wrapped around the leg wire. As you can see, the edges of the skewer slot are rolled over to make them smooth and to reinforce the edges of the slot.

■ *Candlestick,* Joseph A. Messersmith
Mr. Messersmith (left) was 94 years old when he made the traditional candlestick shown below. He continues to make replicas of historical tin items.

■ To form the seam, Peter uses a burring machine. This narrow 90° bend will be folded over a similar bend on the piece that joins it, and the seam will then be soldered.

■ *Flowers* (above and right), **Lee Carter**
Lee designs and produces these flowers in Mexico, with the help of a wonderful family, and he sells the work from a showroom in San Francisco. Lee is committed to using recycled tin in many of his products, not only because doing so is friendly to the environment, but also because the material makes for exceptionally interesting designs. The family salvages rejected sheets of tin that were originally destined to become beer cans or batteries.

■ *Sunflowers*
Dick Crenson
Dick's wife bought a carload of discontinued nail-polish colors at a church bazaar, and he used some of them to paint these cat-food-can flowers. Compare them to the Mexican flowers on this page, with their random printed patterns.

JOSE DE CREEFT

The ambition to produce in quantity is dangerous because it deprives you of the enjoyment of being both a friend to your ideas and a friend to your materials. Materials have qualities which will speak if you listen. Let the artist go slowly; he must give time and labor without counting.

■ *Coffee Time* (left), Laurie Flannery
This Jangled Jumping Jack Clock Man was made in Charm City (Baltimore, Maryland), where the artist lives.

■ *Ring and Bra* (above and right)
Sharon Wasserman
The flowers in these pieces were made from Arizona Iced Tea cans. The superstructure of the bra is a pair of funnels, and everything is held together by strips made from the artificial grass used on athletic fields.

CHAPTER SIX:

PATTERNS

■ *Captain's Lantern,* Richard Haddick
You'll find additional information and more photos of this
lantern on page 76.

MAKING A pattern for a cylinder shape (see the Can-ary Bird Whistles on the opposite page), is fairly easy. Calculate the length and the circumference of the cylinder you'd like to create. Then, on a sheet of paper, draw a rectangle with these dimensions, and you'll have the pattern you need. Cut the pattern out with scissors and trace it onto the tin. (You may want to fasten the pattern to the tin with rubber cement or clear tape and use it as a cutting guide instead.)

If you plan on making several items of the same shape, take a tip from professional tinsmiths: Make a metal pattern. Use a sharp scriber or an awl to scratch the pattern outline onto your material, cut out the shape, and you'll have a pattern that can be reused indefinitely.

Figuring out how to cut a piece of metal so it will roll up into a cone is a bit harder, but easy enough to learn. To reproduce an existing cone shape, such as a funnel, just tape a piece of paper in place around the shape, and trim it to size with scissors. To make a cone-shaped pattern based on a drawing or photo, you'll need a piece of graph paper, a ruler, and a caliper-type compass.

First, refer to the photographed grid on this page. Following the grid lines on the paper, draw a line straight down the middle, stopping 1" (2.5 cm) from the bottom. Divide your desired cone's largest diameter by two to find the radius. Draw line A-B to this length.

Next, decide how high you'd like the cone to be. Measure that distance on the graph to mark point C. If your cone will have an open end, mark the radius of that opening as line C-D. With a ruler, connect point B and point D, continuing the line up until you cross the vertical line at point E.

Set your compass to the length of E-B and draw a half circle, using E as the center. Set your compass at length E-D and using E as a center again, draw another half circle. Now set your compass at length A-B and mark a total of six intersections around the semicircle. Draw a line F-E from the uppermost intersection point on each side. To add enough metal for an overlapping solder seam, just draw a line parallel to line F-E (see left-hand side of photo).

If the cone you're making will come to a point at the top, line C-D won't exist; the pointed end of the cone (or apex) will be point E. With a little practice, drawing and cutting out paper cones of different sizes will soon be very easy.

■ *Miniature Water Pump*
Artist unknown
Dan Van Allen found this miniature water pump in India. It really works.

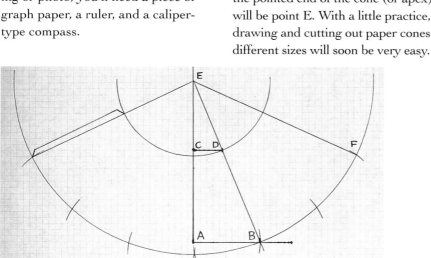

■ *Cone Pattern Diagram*

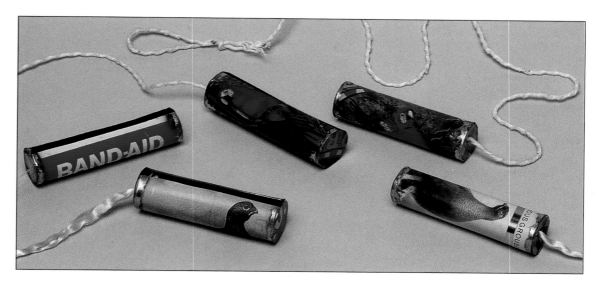

■ *Can-ary Bird Whistle*
Bobby Hansson

■ *Antique Funnels,* **Artist unknown**
Both funnels belong to Betty Oliver.

CAN-ARY WHISTLE

You'll amaze your friends when you demonstrate this can-ary bird whistle, which produces a melodious series of whistles and peeps when it's swung around by its cord. This model was made from a 1-1/2" x 2" (3.8 x 5.1 cm) tin can section and two 3/4" (1.9 cm) tin can squares.

Start shaping the cylinder by bending about 1/8" (3 mm) of each long edge of the larger piece to an angle of roughly 5°. The easiest way to do this is to let the edge hang over the edge of a cutting board and use a mallet to tap it lightly, moving from one end to the other. Pre-bending these edges before bending the whole strip into a cylinder is important; it compensates for the fact that metal has a "memory" and wants to spring back to its original shape. If you form the cylinder without pre-bending these edges, you'll have a very hard time getting them to conform to the cylinder's shape.

Next, shape the metal around a 1/2" (1.3 cm) dowel. Start with one edge and roll toward the middle. Then repeat from the other edge. You'll

probably be able to do this with your fingers, but if not, don't despair. Tap the metal lightly with a small mallet.

Take the rolled metal off the dowel and pinch it to close the cylinder a bit, being careful to retain the round shape. You're aiming for a cylinder with a 1/8" (3 mm) gap between its edges, as shown in the photo. (The gap is to let the peeps out!)

Next, close the ends of the cylinder by soldering the square pieces onto them. Make sure to make these nice and tight. Then trim the end pieces until they're round and file any rough edges to smooth them. (Shaping a clean joint this way is easier than cutting a sized circle and trying to solder it for a perfect fit.)

Make a small hole in the middle of one end piece and push one end of a piece of cord into the hole. Fish the cord out through the peep slot, tie a knot in it, and tug it back until it stops. Cut the cord about 2' (61 cm) long, tie a finger loop in the end, and swing the whistle around in a circle; it should produce a series of intermittent chirps. For optimum results, you may need to adjust the whistle a bit. Try experimenting with different sized cylinders and different kinds of string. Look for cans with printed pictures of birds or appropriate words. Cheep trills!

Nutmeg Grater

One of the best ways to learn a skill is to find something that appeals to you, take it to your workshop, and try to copy it. Making a copy probably won't be as easy as you'd hoped, but if you keep at it, you'll learn something. You might even work out some variations to improve the original design. Make several copies and as many variations as you can think of.

I certainly don't advocate forgetting where the design came from or selling copies of other people's art. Doing so would be unfair at best — and, at worst, illegal. Every artist worth his or her salt, however, started out by copying the work of someone more skilled. As you become more familiar with tools and techniques, you'll find it much easier to develop your own designs and ideas, but in the meantime, judging your work against a model makes it possible to see where you've succeeded and where you haven't. Then you can try to modify your shortcomings on the next attempt.

These little nutmeg graters are very handy and make great gifts. I borrowed one from my wife so I could figure out how to make my own version, and I came up with the following method.

The first step is to make a pattern from the grater you've found. Set the grater flat on a piece of fairly stiff

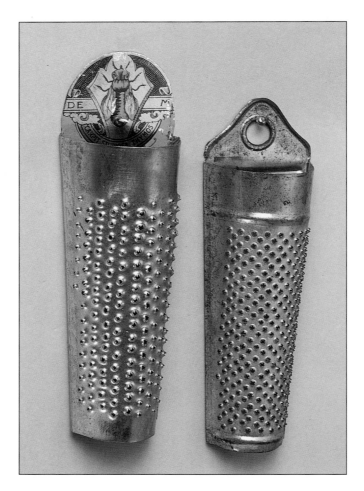

■ *Bug Grater* (and model) Bobby Hanson This bug grater hangs by a nail that also forms the bug's stinger.

tracing paper. You'll notice in the photo (right) that my grater is made from a single sheet of tin with a seam tab on each side. Trace the outline of the flat side of the grater, but on one side, draw the line about 1/4" (6 mm) out to allow for a seam tab.

Use scissors to cut along the seam line and the two shorter lines. Then place the grater back down on the cut pattern. Bend the pattern's seam tab up and tape it to the inside of the grater. Now wrap the tracing paper snugly around the curved front of the grater and finish tracing the grater's outline. Add about 3/16" (4.5 mm) along the other seamed side to make another seam tab. Remove the traced pattern and cut it out.

I found an olive-oil can with a cute bug picture on it and centered my pattern so the bug would be in the middle of the curved upper section. Use rubber cement to fasten your pattern to your tin. (Transparent tape will also work and is a better choice if you want to reuse the pattern.)

Cut around the pattern's outline. Remove the pattern and place the tin, with what will be its inner surface facing up, on a surface such as a piece of old inner tube. Scratch some guide lines for placement of the grating holes. Then, using a hammer and a sharpened nail,

punch the holes which make up the grating surface. (Practicing on a scrap of tin will help you determine how hard to hit the nail.)

Clamp what will be the flat back section of the grater in a vise and make a 90° bend. Remove the bent tin from the vise, and use a dowel to form the curved front surface of the grater. (Place a piece of inner tube over the punched holes to protect them—and your hands.) You'll need to fiddle around a little to get the curve to come out right and to make sure everything fits. The secret is to go slowly, bending just a little of the tin at a time. If you try

to form the curve with one whack, you'll probably end up with kinks and creases in the tin; these are very hard to repair.

When you're finished, bend the tab on the curved section back on itself and make a double bend on the flat end, as shown. Now hook the tabs together and pinch the seam shut with needle-nose pliers. Until you get the hang of it, this last operation entails quite a bit of tinkering. Don't get discouraged! You may want to modify the design of this seam by soldering it instead of crimping it. If you do, be sure to use a lead-free solder.

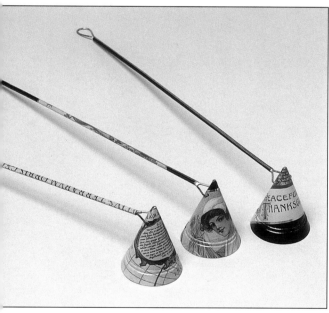

■ *Tinworks,* Richard Haddick
Richard is obsessed (appropriately) with proper placement of printed patterns. As a true tinsmith of the Old School, he doesn't leave anything to chance. When placement of tin can print or imagery on your finished piece is important and you're working with a shape such as a cone, make a pattern from fairly stiff but flexible clear acetate or tracing paper, both of which can be formed into shapes. Using these flexible patterns will help you line up the design exactly.

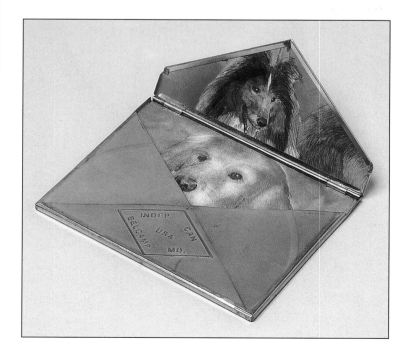

■ *Captain's Lantern* (top left and above)
Richard Haddick

This lantern (also shown on page 71) is a copy of the type that used to hang in the masters' cabins on the tall ships of long ago. The beautiful old map was printed on a wastebasket-sized pop-corn can. Richard gave the popcorn to his pet geese and made this master-piece from the can. Notice the exquis-itely careful placement of the printed design on the lantern.

■ *Ham Megaphone* (left), Pam Lins
A silver handle proves to be the ultimate accessory for this charming megaphone. Pam made it, of course, from a ham can.

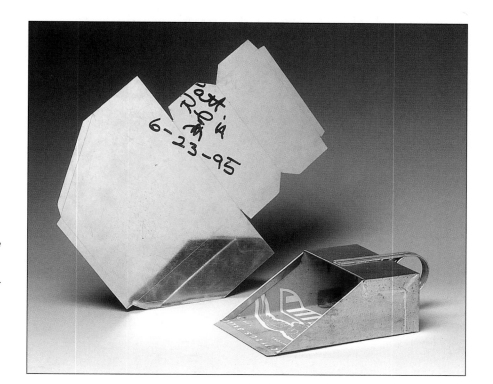

■ Patterns

To help figure out how to make a shape such as a box, draw the shape on thin cardboard (cat-food boxes or old file folders work well). Cut and fold the pattern, and tape it together to make sure everything looks right before you cut the metal. It's easy to trim the cardboard or add extra sections as necessary. When you're ready use the pattern, just flatten it out and trace the design on the tin.

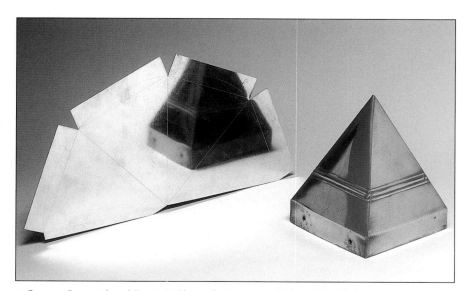

■ Copper Pyramid and Patriotic Flagpole Protector, Richard Haddick
The Copper Pyramid was created to protect the top of a 4 x 4 mailbox post. The pattern, which stands behind the finished piece, was also used to make the flagpole protector.

■ *African Briefcase*
Collection of W.H. Bailey and Laurie B. Eichengreen
Holy Mackerel! What a briefcase!

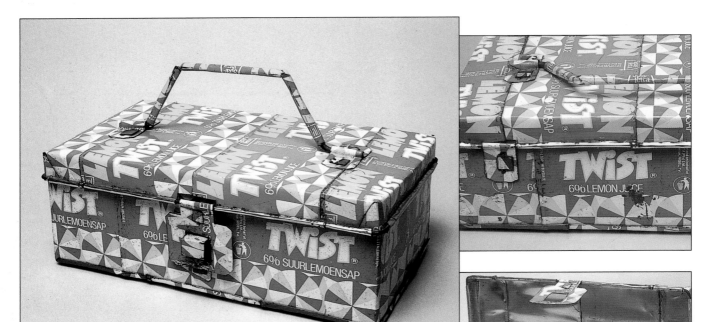

■ *Green Lidded Box*
Collection of W.H. Bailey and Laurie B. Eichengreen
Appreciate the charm of this piece, by all means, but take a good look at its ingenious hardware construction, too.

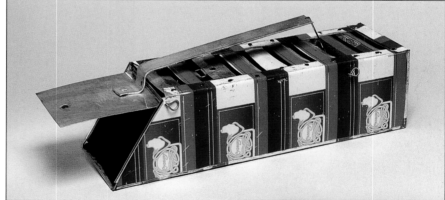

■ *Briefcases,* Artists unknown

Spin Something, in New York City, imports and markets these beautifully designed briefcases from India. The artisans obtain misprinted sheets from tin factories and turn them into luggage. Although these items are mass-produced and include store-bought hardware, they make excellent design textbooks. The slightly curved shape is extremely strong as well as attractive, and the reinforcing members on the interior surfaces add considerable stoutness without adding much weight.

■ *Rat Trap,* Artists unknown

Just as the briefcases on these pages are ideal for the fledgling entrant in the "rat race," this Indian-made rat trap, imported by Homefront (Hillsborough, NC) is the best rat catcher I've ever seen. My wife, Maggie, and I put one in our barn and caught five rats in six days. I was born in the Year of the Rat and am superstitious about killing these rodents, so I'm pleased to report that when the rats were released, they were very perky.

Take a look at the homemade rivets in the close-up. They're made from sawed-off nails.

■ *Tin Altars*
Top left: Collection of W.H. Bailey and Laurie B. Eichengreen
Above: Collection of Lee Carter
The author guarantees that if you study and compare these icons for five minutes, you will be a better person. The icon on the left is the older of the two.

■ *Como un Recuerdo* (left)
Bobby Hansson
Ann Parker and Avon Neal bought this backdrop in South America and let me use it to set off a photo of a dear friend who loved Mexico. I made this little shrine of cans and bottle caps after seeing a collection of Mexican tinwork.

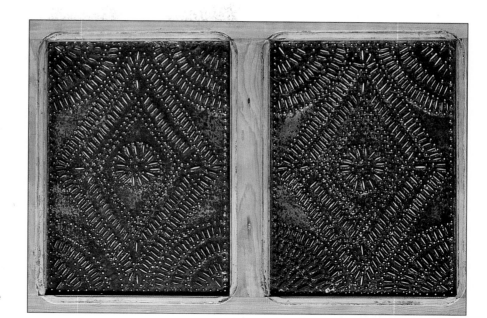

■ *Pie-Safe Panel,* Artist unknown
The decorative punched holes in this pie-safe panel provide ventilation, while the panel itself keeps out flies.

To punch holes in tin, a pattern is carefully positioned and taped to the tin sheet, which is held down on a wooden board. The holes are then made with a pointed round punch and a straight, chisel-like punch. The piece shown below the pie-safe panels will be bent in a circle to form a lantern body. The pattern formed by the holes adds visual interest, provides air for the candle, and prevents strong drafts from extinguishing the flame.

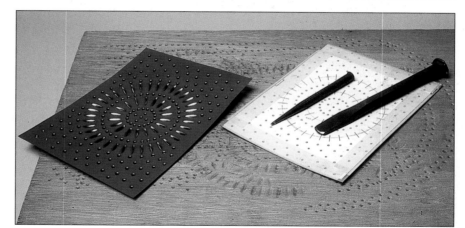

■ *Mexican Watering Can*
Artist unknown
You will never see or need a nobler watering can than this classic of Lee Carter's from Mexico.

■ *Two Lanterns* (top left)
A traditional pierced lantern stands cheek by jowl with a tin can reproduction by Richard Haddick.

■ *Lantern* (above), Richard Haddick
This tin lantern includes a customized Rotary logo.

■ *Lantern* (left), Richard Haddick
The artist made this lantern from a Christmas-cookie tin.

CHAPTER SEVEN:
TOYS

■ *Tugboat,* Bobby Hansson
Fashioned for Popeye the Sailor, this squat tugboat was made from an Olive Oyl can. After some judicious cutting, bending, and riveting, I added a luncheon meat can as a wheelhouse; no square spinach cans were available. A hair-spray smokestack, a renowned celestial navigator at the wheel, and a cry of "Avast ye swabs!"—and it was launching time.

THE ROUND ends of cans look so much like wheels, and square cans placed on their sides look so much like buses or train cars, that it's almost impossible to resist the urge to turn cans into playthings. Much of the fun is in trying to create a believable truck or airplane in which the original can can still be recognized. Sometimes it's a challenge to see how little you can do and still make the transformation work. At other times, it's enjoyable to change the original can so much that it can't be seen in the final toy unless it's pointed out. Series of toys that are similar but not identical, such as the *Spam Can Tank* on page 35, are also amusing to make.

In case it's not obvious, I should point out that tin can toys are for adults and for children old enough to handle them. Don't offer these creations to toddlers. Citizens who tend to stick everything possible into their mouths should not have access to these toys! And be sure to round all sharp corners on tin can playthings, no matter who will be doing the playing.

■ *Pickup Truck,* Bobby Hansson
This little pickup truck, with its driver and passengers glued in place, was cut out of a cracker tin. Use a stiff piece of paper to work out the design—a process that's actually fun—and either tape or glue it to the metal to serve as a cutting pattern. After cutting the tin, fold it and solder the seams.

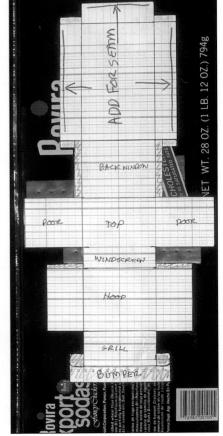

■ *Airplane,* Bobby Hansson
I cut and folded the airplane body from a single flat sheet. The wing was made with another flat piece that was bent around a dowel to form the curve. After I soldered the wing and body together, I spray-painted the plane silver.

■ *Truck,* Bobby Hansson
As this cocoa-powered truck races down the highway, we can see the driver from the front and from the side. To make wheels and axles, I used nibbling shears to cut the tops from aerosol cans. (These shears are expensive but are ideal for cutting around the sidewall of a can because the waste strip curls up between the top blades and keeps the shears from binding.) Then I flattened the can tops and pushed nails through the spray holes. I soldered the nail heads to the caps and inserted the nail ends into a rolled-up tin sleeve, which was soldered to form a solid axle.

■ *Bus,* Collection of W.H. Bailey and Laurie B. Eichengreen

Before the development of plastics, many toys, especially mechanical ones such as trains, planes, trucks, and cars, were made of tin. They were light, inexpensive, and relatively sturdy, and were painted with bright colors and many realistic details. In countries where commercial tin toys weren't common, people often used tin cans to make their own toys. It's a shame that when countries modernize, this kind of traditional self-sufficiency tends to die away. Several of the artists in this book incorporate into their work old tin toys that they find at yard sales and flea markets.

The wooden frame of this bus adds weight and stability and provides an easy way to attach the novel suspension system. The axle goes through the wooden wheels and is bent back like a staple to fasten the wheels securely.

The bus—and the trucks on the opposite page—are all handmade, and most make use of tin from tin cans. The trucks were made for personal use and were not intended, as are most similar objects today, for sale as souvenirs.

■ *Beer Truck,* Collection of W.H. Bailey and Laurie B. Eichengreen
The frame of this beer truck is made of tin that is bent at right angles for extra strength. The axles are soldered in place, and jar lids serve as wheels. Note the sewn canvas load cover, the radio antenna, and the side mirror.

■ *Noise-Maker Truck,* Collection of W.H. Bailey and Laurie B. Eichengreen
This beautifully detailed truck, which functions as a noise maker during festivals, is probably a model of one owned by its creator. When the handle is grasped and the truck twirled around, the ratchet on the bottom is snapped loudly against the wooden strips located underneath the truck body.

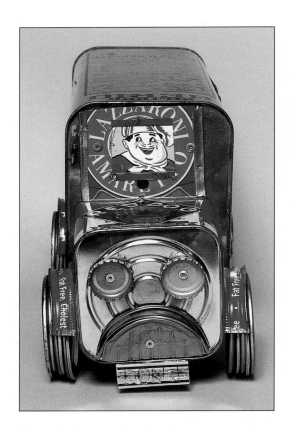

■ *Don Pepino Truck*
Bobby Hansson
Take a look at the red motor compartment in the photo above. It shows the top flap bent up to form part of the windshield and the bottom flap trimmed and drilled. This lower flap fits into a slot cut in the oil can with a knife blade, and a pop rivet holds it in place. Two rivets go through the side sections of the windshield, as shown in the photo on the left.

■ *Trucks,* Bobby Hansson
Compare these two trucks. They're both based on the same general design, but they have slightly different grills, tires, and axles. When you make your own truck, don't be afraid to change or improve the design.

■ *Don Pepino Truck,* **Bobby Hansson**

BOBBY HANSSON

I used to love the black-and-white Italian movies of the '50s and '60s and decided to build an Italian truck as I remembered them from those films. I started with an olive-oil can, which seemed appropriate because it was a "Prodotto Italiano." The engine compartment was cut from an Italian cookie can. Jar lids cut in half became fenders. As I worked, I conjured up scenes based on those great old films.

It's a dark and stormy night. The tall, narrow truck looks dangerously top-heavy, moving down the twisting mountain road. In the cab, we see through the intermittent wiper a stubble-jawed man struggling with the big steering wheel, a cigarette dangling from his lips. He is squinting as he tries to see the dimly lit road.

We flash to the kitchen back home, where a woman wearing a black slip is ironing—Sophia Loren, perhaps. In the background is an older woman in an apron-covered dress, stirring something on the stove. She speaks, and the blurry white letters across the screen say, "I sure hope Frankie is gonna be OK. He needs new tires on that truck."

The young wife tries to be brave. "Don't worry, Mama Giorgini. He's a good driver; he'll be all right."

Back to the truck, to the "brrrrr" of the motor on the sound track. Now a close-up of the front tire, splashing through puddles, its woven white cords showing through the rubber sidewall. Frankie takes a deep drag on his cigarette, while the steering wheel jiggles like something alive in his big-knuckled hands.

"Brrrrr." The truck is swaying as the stormy wind buffets its tall sides. A branch blows down, and the weak tire rolls over it. Are those white cords bulging more than they were? It's hard to tell in the dark; clouds keep drifting across the moon.

We return to the kitchen. "I wonder if he's there yet. If he doesn't get that load to the market on time..."

A huge black cloud starts across the moon. "Brrrrrrrr, brrrrrrrr."

I hear a voice asking, "Are you all right?" It's my wife. "I heard growling." Then she sees the truck in my hands and guesses what's happening. "You've been in here six hours. Are you hungry?"

"Do we have any spaghetti?" I ask. "I need a Don Pepino can."

O.K. Maybe I'm a little silly, but I really have fun working with tin, and time does fly by. The next morning, as I made another truck for my Italian fleet, I played a recording of Fellini sound tracks.

■ *African Trucks* (above)
Collection of W.H. Bailey and Laurie B. Eichengreen
These tiny trucks are made from folded strips of tin.

■ *Javanese Truck*
Collection of W.H. Bailey and Laurie B. Eichengreen

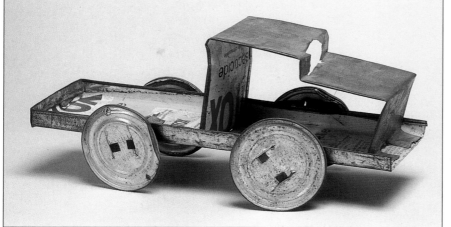

■ *African Truck #2*
Collection of Bobby Hansson
This crude but charming African truck was made from an insecticide can.

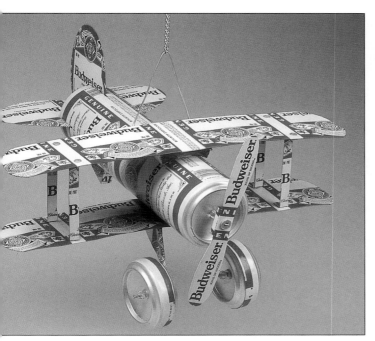

■ *Brew Bird of Happiness* (1995), **Peter Riegert**

■ *Beer-Can Plane,* **Joseph Bomba**
Noteworthy for its neat and careful construction, this plane
looks as if it's really flying.

PETER RIEGERT

*I was born in New York City and
now live in California. I'm a believ-
er in recycling and have always
tried to breathe new life into dis-
cards. In the case of the "Brew
Bird of Happiness," half the fun
was in obtaining the raw materials
for the piece.*

■ *Tin Can Chopper from Mali,* **Collection of Craig Nutt**
Spray-can wheels and a skeeter on the seat distinguish this racy road toy.

■ *Airplane,* **Artist unknown**
Lynda Curtis bought this toy airplane in
Africa. Because it's a push toy, it has
big wide wheels and a long wire handle
attached to its tail. The heavy wire
superstructure on the bottom, combined with an oil-can "skin," makes this
plane quite sturdy and durable.

■ *Watering Can and Pitcher*
Collection of Bobby Hansson

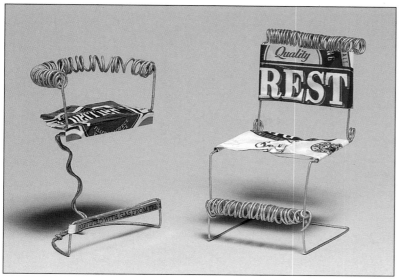

■ *Chairs: Perrier & Quality Rest (1993),* **Julie Flanigan**

■ *Tiny Chair,* **Janet Cooper**

■ *Rosie the Riveter and The Skinny Elvis*
Sue Eyet

The twentieth-century icons pictured here are based on the dancing figures which are common folk toys in the rural South. In use, a thin board is held under the operator's thigh and thumped rhythmically with a fist while the articulated figure is held in light contact with the bouncing board. The vibratory contact imparts a lifelike dancing motion to the puppet.

The retail outlet of the Independent Can Company (Belcamp, MD) sells seconds and overruns of the company's commemorative cans. Sue bought an Elvis can and a reproduction of a cola container with a comely lass on it. She cut out the faces and taped them down to her drawing board. Next, she sketched bodies in the scale and spirit of the faces. Then she traced the drawings onto plywood and cut the shapes out with a jigsaw.

Rosie's hat, shirt, and jeans were cut from Sue's trove of cookie cans. The tin pieces were fastened to the plywood with tiny nails. Sue made and inserted rings to hold the articulated parts together. The working versions of these dancing figures (Sue's don't dance) have joints made of interlocking eyehooks. To make the spangles on Elvis's suit, Sue used glass beads and tiny lock washers. Other details, such as Rosie's bunch of keys and Elvis's hair and belt buckle, add a humorous touch.

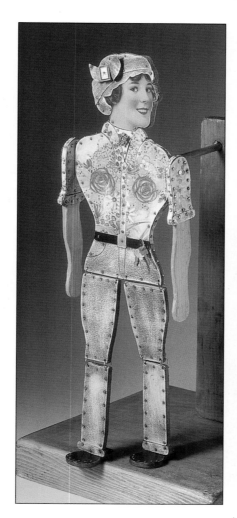

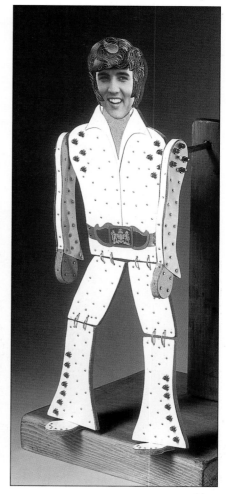

LETTER STAMPS

Steel letter stamps add some fun to tin work. I bought one set at an auction and another two—one with reverse letters—from M.S.C. Industrial Supply Company (see "Suppliers" on page 144), and I love to use them to personalize badges, awards, and other projects. Most machinist's-supply stores carry stamps in a variety of sizes.

The photos on this page show some projects I made for some pals involved in recycling. According to my pals, Giuseppe and Ada, the inscription *Io amo la immondizia* means "I love recyclable trash."

You may find it hard to keep these stamps straight and evenly spaced, so practice on scraps until you get the hang of using them. Support the metal on a heavy block of wood as you work, experimenting with different blocks until you get the results you want. Soft wood makes letters less crisp-looking than hard wood, but if the wood is too hard, the letters won't be deep enough to read easily. If you hit the stamp too hard, you may cut right through the tin; in some cases, this actually looks good.

Reverse letters should be stamped from the back of the piece, so they'll be raised when viewed from the front. Flip to the photo of Harvey Crabclaw's frame on page 58. To achieve this effect, first stamp the metal from the back. Then attach some emery paper to a flat block of

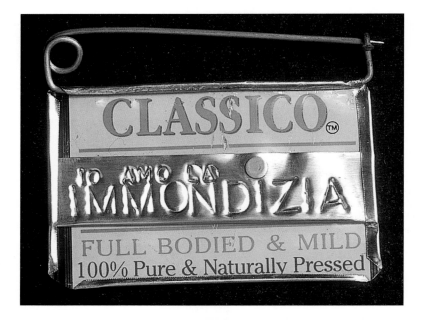

■ *Badges and Awards,* **Bobby Hansson**

wood and use the paper to sand the raised letters down to bare metal.

Pin backs and earring hardware are available at many hobby and bead shops. Attach them to your badges with rivets, small bolts, or epoxy glue. For my own recycled awards, I make big diaper pins from old bicycle spokes. I bend the spokes with needle-nose pliers and sharpen them with a small file.

Harvey Crabclaw used a few of my letter stamps as abstract design elements in some of his picture frames. The photo on page 13, for example, is of a woman whose initials are M.M. To emboss the frame he made, Harvey punched M's to run in every direction.

■ *Get Well Can*
David Mazzarella
The nurses were amazed when this tin collage with stamped letters arrived in my hospital room. "Someone must really want you to get better," one of them exclaimed.

■ *Bus,* Collection of Elizabeth Seltzer
Elizabeth found this flame-cut bus in the Caribbean. The bright paint job adds to its charm. The hook on it may be for a bus ignition key.

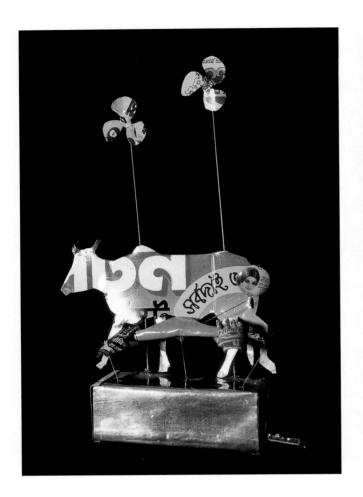

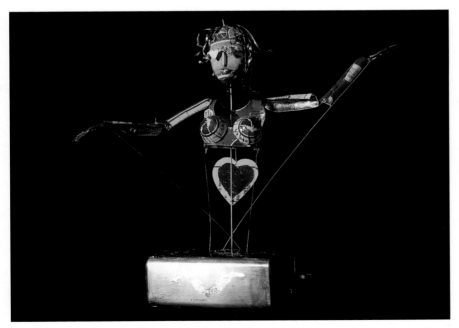

■ Annabel Hill, a 3-D illustrator who lives in London, made these hand-cranked automata after a trip to India. They illustrate different characteristics of Indian life as seen through a tourist's eyes.

Clockwise from top left:

Motor Cow shows motor vehicles and cows on Indian streets, all going in different directions.

Tourist Looking depicts a group of people staring as if they had nothing better to do. One of the eyes winks when the handle is turned.

Madhuri Dixit is a dancing film artist.

CHAPTER EIGHT:
MUSICAL INSTRUMENTS

■ *The Gitana*, Bobby Hansson
This instrument is geometrically
designed for square-dance music.

LORRIE GOULET, Jose de Creeft's wife, was asked in 1966 to appear on a WCBS-TV children's educational program to talk about art. I went along to take photos of her TV debut. Lorrie mentioned my musical instruments to the producer, and before I knew it, I was a regular guest on the show.

As I made a musical instrument out of a tin can, I would explain how and why it worked. Then Mollie Scott, the hostess of the show, would come out with her guitar, and the children would join us in singing a folk song. I made enough instruments for everyone, and we had a lot of fun while the children learned about sound.

One way to change the note a plucked string makes is to tighten or loosen the tension on the string, as you do when tuning a guitar. The washtub bass fiddle, for example, is a folk instrument made from a washtub turned upside down, with a string attached in the center of its bottom. The other end of the string is tied to a broomstick held at the rim of the tub. The player puts a foot on the rim to hold everything in place and by bending the stick back and forth, changes the tension of the string, which changes the notes. The tighter the string is, the higher the note. The washtub acts as a resonator to make the sound louder and also affects the sound's quality.

I made my version of the washtub bass fiddle by using a #10 tin can, an old tennis racket my brother gave me (with no strings attached), some baling wire, a nail, and a little metal cap from a spray-paint can. The only tool I used was a pair of pliers. I made three holes in the can using the nail as a punch and the pliers as a hammer. I wired the can in the middle of the racket, threaded an old guitar string through the hole in the top of the can, drove the nail into the end of the racket handle, and tied the wire to the nail. I dented the paint-can lid and wedged it in under the string to position it.

To play this instrument, you hold the racket head between your knees. When you squeeze your legs together, the oval shape gets longer

■ *The Tennis Racket,* **Bobby Hansson**
Use this instrument to play a score you love; the net result will serve you right.

■ *Tin Can Alley Sign,* **Sue Eyet**

and the note gets higher. Make a racket with your racket, and develop legs like Martina Navratilova's while you do.

We made another one-string twanger out of a golf club, using a much smaller can just to see what difference its size would make in the sound. The *Hot Iron* is played by bending the shaft of the club.

To show that shortening the string changes the note, we made another instrument from a hockey stick with frets created from coat-hanger-wire loops; these frets are similar to the ones on a guitar. We called it a *Hockey Pluck*. The similar *Field Hockey Schtick* is played by holding a piece of iron firmly against the string to "stop" it in different places.

■ *Hot Iron,* **Bobby Hansson**
Now here's a can you can swing with at nightclubs, when you're teed off and not up to par.

■ *The Hockey Pluck,* **Bobby Hansson**
With its bent coat-hanger frets, the Hockey Pluck produces specific notes.

■ *The Field Hockey Schtick,* **Bobby Hansson**
This instrument is ideal for swinging at balls. A handheld object changes the length of string that vibrates, thus changing the note.

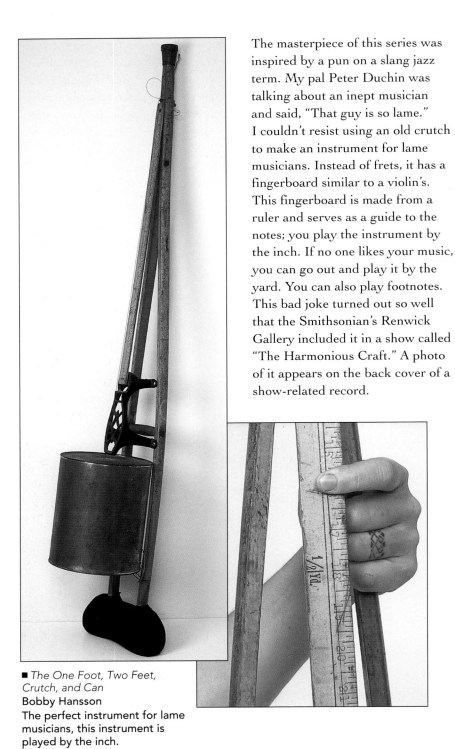

The masterpiece of this series was inspired by a pun on a slang jazz term. My pal Peter Duchin was talking about an inept musician and said, "That guy is so lame." I couldn't resist using an old crutch to make an instrument for lame musicians. Instead of frets, it has a fingerboard similar to a violin's. This fingerboard is made from a ruler and serves as a guide to the notes; you play the instrument by the inch. If no one likes your music, you can go out and play it by the yard. You can also play footnotes. This bad joke turned out so well that the Smithsonian's Renwick Gallery included it in a show called "The Harmonious Craft." A photo of it appears on the back cover of a show-related record.

■ *The One Foot, Two Feet, Crutch, and Can*
Bobby Hansson
The perfect instrument for lame musicians, this instrument is played by the inch.

■ *Citizen's Cane,* Bobby Hansson
Here's an instrument for the slightly lame musician who begs to differ. When street musicians seek alms, the swagman hobbles back and forth in front of the other musicians. By squeezing the handle on the peanut can and moving it across the ball-point plucker, a piteous twanging sound is produced. The musician may want to call attention to the can label and claim to be "working for peanuts." He might also explain that when the tin-can resonator is filled with spare change, it isn't as loud.

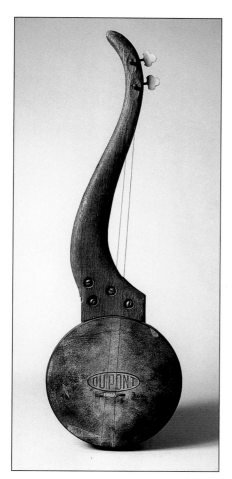

■ *The DuPont Baby Sitar,* **Bobby Hansson**

■ *Tin Guitar,* **Bobby Hansson**
A distant cousin of the traditional steel guitar, this instrument was made with a salvaged guitar neck.

Inspired by this success, I added a drill to my toolbox and made the *DuPont Baby Sitar* from a 16mm film can and a piece of an old dresser. I boldly used twice as many strings, as well as real tuning pegs from a wrecked guitar.

One night, I found a big box filled with broken instruments in the garbage behind a music store. I used a banjo and a chicken-liver tin lid to make a *Canjo,* and another banjo neck on a soda-cracker can to make a *Gitana* (shown on page 97). A guitar neck attached to a 35mm movie-film can made an almost real guitar. Eric Weissberg, who played the (real) guitar on the soundtrack of the movie *Deliverance,* played a tune on my tin guitar when he was a guest on the TV show, *Around the Corner;* it sounded great!

As you can see in the photos, the tin cans in the single-string instruments have only a few holes drilled in them to accommodate baling wires. The wires help suspend the cans in mid-air by pulling down to balance the force of the sounding strings that pull upward.

The multistringed instruments employ bridges to transmit sound vibrations to the top of the can. The can works well as a resonator, or sounding board, because it is uniformly thin—in fact, very thin. The can is still strong enough, however, to allow you to stretch the guitar string very tight without having it rip through the metal or collapse the can. Using nails to punch holes in cans is very easy, yet the metal around the holes is strong enough to hold the wires. One of the most

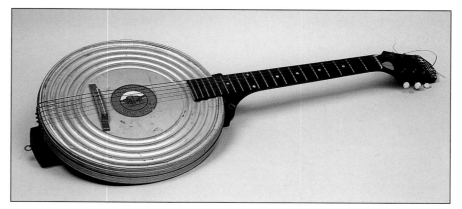

■ *Weaver's Inspected Canjo,* **Bobby Hansson**

exciting qualities of tin is that you can use simple tools to cut, bend, or drill it quite easily and still end up with a very strong object.

You've probably heard of the Juilliard String Quartet. With a dozen tin cans and using tools you probably own already, you can create a *Junkyard String Orchestra*. You might even compose a cantata for cans. (Turn to pages 47 and 73 for instructions on making two kinds of tin whistles.) The last time I demonstrated these instruments, we sang the following lyrics to the tune of "Amazing Grace."

> I'm saving waste
> That I have found.
> It's useful as can be.
> What some call trash
> Is valuable
> If you can learn to see.
>
> Once garbage filled my heart
>   with fear
> But now my sight is keen
> How precious does that waste
>   appear
> Thanks to recycling.

You have to get the accent just right to rhyme "keen" with "recycling."

■ *Folgerphone,* Craig Nutt
This woodwind—or rather tinwind—has a saxophone mouthpiece, a copper-pipe fingering part, and a Folgers coffee-can body. "It's Mountain Grown—that's the richest kind," according to Mrs. Olsen. This Folgerphone is featured on the record *Dinosaur Time,* which Say Day-Bew Records released in 1980.

■ *Tin Can Symphony,* Bobby Hansson
I wired three potato-chip can lids to a Symphony Ready Rubbed Tobacco tin and mounted the lids on the skeleton of a hot plate. Pulling on the Symphony tin changes the notes by changing the tension on the melody string. The four harp-like strings can be plucked outside or inside the nest of lids to yield different chords.

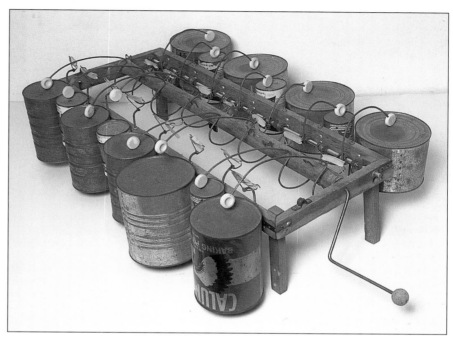

■ *Cranky #1,* Jody Kruskal
This is a gravity powered, hand-cranked, programmable rhythm machine.

■ *Hurdy-Gurdy,* Jody Kruskal
The pedal of this electric harmonic hurdy-gurdy adjusts the tension of the disc bow on the 6'-long (183 cm) strings. The gurdy includes steel strings and a plastic wheel that is driven by an electric motor. Because this instrument has an amplifier, it makes a very loud—almost scary—modern sound.

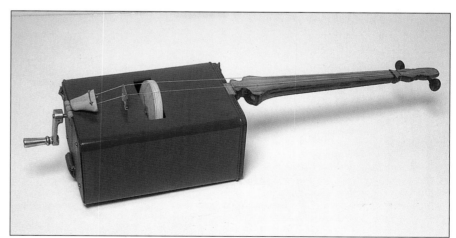

■ *Fiddle-Gurdy,* Jody Kruskal
Made from a large pepper can, this instrument has two strings. It's played by turning the crank handle, which turns a wooden wheel that rubs against the strings as if it were a bow. The difference is that the wheel plays continuously. One string is a drone string, which keeps playing the same note, like a single-note drone pipe on a bagpipe. Jody plays tunes by fingering the other string, the way he would on a violin. When he sings along, the sounds remind me of music from hundreds of years ago; they're very strange and beautiful.

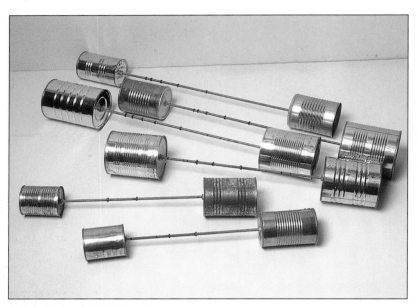

■ *Dumbbells,* Jody Kruskal

The cans operate in two ways: as bells when they're struck and as resonators, which amplify the various harmonics of the aluminum rod joining them when the rod is either struck with a mallet or bowed.

■ *Dumbbells,* Jody Kruskal

■ *Three-Note Can Chimes and Water Gongs,* Jody Kruskal

■ *Marching Can Chimes,* Jody Kruskal

■ *Styrofoam Amplified Can Chimes*
Jody Kruskal

JODY KRUSKAL

I started building unusual musical instruments in 1984 for my work with the Mettawee River Company, a touring puppet theater group lead by Ralph Lee. As composer and musician for these outdoor summer shows, I wanted to extend my palette of available timbres beyond the traditional instruments for which I'd been writing. Homemade instruments provided a simple and effective solution.

We used no amplification, so the instruments had to be loud; their visual impact was an added plus. After five summer shows with Mettawee, I had dozens of string, wind, and percussion instruments on hand, and I formed the Public Works Orchestra in 1989. Since then the orchestra has performed numerous dance scores, concerts, parades, and workshops, using predominately homemade instruments, including the can instruments I've made that are pictured in this book.

Although I designed and built these particular instruments, I was not the inventor of the basic idea. The first time I saw cans used this way was in the mid 1980s at a Bread and Puppet performance at St. Ann's, in Brooklyn, New York. Peter Hamburger had built can-resonated hurdy-gurdies and chimes and used them beautifully in Peter Shuman's Washerwoman Nativity.

Can instruments appeal to me for several reasons. They sound great, with their harmonically rich, distinctive tones. They look intriguing, too—the old "silk purse out of a sow's ear" idea. Cans are versatile. You can scrape their ridges or fill them with beans and shake them. Striking their bottoms produces a pitched clang. Playing many cans together makes a steel drum sound, of sorts. Hang a can by its edges and put in a little water to create a tunable gong. Use the bottom as a membrane for amplifying strings like a banjo or harp. Make can chimes by attaching a long bolt to the bottom membrane. Scraping or striking the bolt produces a variety of sounds. Pitch is determined primarily by the length of the bolt, with a range of over three octaves.

Tin can instruments can be quite loud because cans are efficient sound resonators. This is due to their low ratio of weight to surface area, the springiness and conductivity of thin metal, and the resonant cavity inherent in a can's shape.

For the most part, can instruments are simple to build. Recycling has made cans readily available in a large variety of shapes, materials, and constructions. Musical instruments have always been built using the materials available to us. The human race has grown from the ancient musical technologies of stone and bone and of sinew and fiber to a high level of craftsmanship and a wide range of materials including wood, metal, plastic, and electronic devices. But if you want to build a musical instrument today, using materials from your home and local hardware store, start with a can.

■ Rattle
Collection of Dan Van Allen
People all over the world
make rattles from materials
such as gourds and turtle
shells. Dan found this tin
rattle in Mexico.

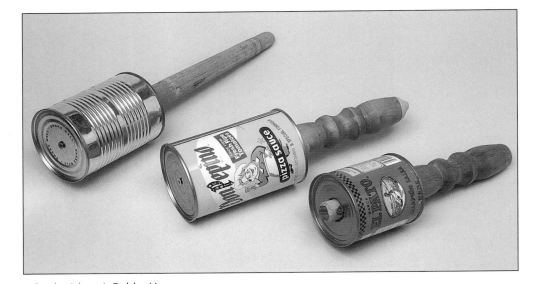

■ Rattles (above), Bobby Hansson
Open the can by poking a hole in the center of the lid. Empty the can, rinse it well, and insert a
dozen pebbles through the hole. (Experiment with changing the sound by using different sized
pebbles and cans). To insert the handle, fit it into the hole. You can push it through another
hole in the other end of the can and let it peek out, or you can jab a screw down through the
other end of the can and into the end of the handle, as shown in the photo.

These rattles have wooden beads, rubber pencil erasers, and small scraps of tin in them.

■ The Tinhorn
Bobby Hansson
One of my favorites,
this is a wind instru-
ment with a bugle
mouthpiece. It
would qualify for the
brass section except
for the fact that it's
made of tin cans
and plastic.

CHAPTER NINE:
JEWELRY, SCULPTURE, AND MORE

■ *Secret Mission (1963),* **Tony Berlant**
11" x 7-1/2" x 5-3/8" (27.9 x 19.1 x 13.7 cm)

THE PHOTOS in this chapter are of one-of-a-kind jewelry and artwork. Revel in them, by all means, but don't, please, copy any of these exquisite pieces. Study the photos carefully. Can you see the beauty in a rusty old scrap of metal ravaged by the sun and rain? Looking closely at ordinary things—leaves, rocks, scraps of wood and metal, the things we ordinarily either take for granted or even shun—helps me see beauty everywhere.

Not everyone wants to wear the kind of jewelry shown in this chapter. You might not be able to figure out what some of it is for, but take my word for it, the people who made these pieces are artists, and the time that you spend looking at and trying to understand their work will be good for you.

RUST NEVER SLEEPS

The primary reason for tinning steel cans is to inhibit oxidation (or rusting). Tin, which like gold is not affected by the combined attack of water and oxygen, protects the vulnerable steel beneath it. In a fully sealed can, there is virtually no oxygen so the steel is safe. An opened can, however, is exposed to atmospheric attack from other gases as well as to corrosive elements such as salt and acid rain, and will develop rust.

Sometimes the rust will cover the entire steel surface, and sometimes textured pitting will occur. When your concern is the safe preservation of food, you'll naturally see rust as a problem. When your goals include the discovery of accidental beauty, however, oxidation is a gift from God. Rust celebrates the multitudinous variations and combinations of natural phenomena, and the circle of existence that encompasses decay and rebirth.

■ *Pendant,* Ruth Abraham Fisher
Found objects, with rusty can part

■ *Iron Oxide Corrosion Product*
Tim McCreight
Ever the metalworking pro, Tim named this pin for the patina he used on it.

■ *Way Ahead of Its Times Brooch (1968)*
Professor Robert Ebendorf
Metal, photo, heel tap, Indian head
penny, tin can top, and pull

■ *The Lady From Norway (1995)*
Teri Blond
Collection of Professor Robert Ebendorf
Norwegian sardine tin and Mexican
metal figure

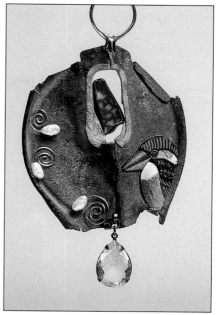

■ *Necklace (1995)*
Professor Robert Ebendorf
Sterling silver, English china fragment,
coat-hanger wire, old gold ring, 18-
karat gold wire, pearls, glass, bird from
tin can, and top of old tin can found on
the street

■ *Cross/Necklace I and Cross/Necklace
II (1993)*
Professor Robert Ebendorf
Cross I: Coat-hanger wire, turquoise,
amethyst, glass, rock and street stones,
pieces from olive-oil can and gas can
Cross II: Coat-hanger wire, olive-oil can,
found parts, mixed media, and wood

Both crosses were made from the
same can, but a variety of techniques
and additions make them quite different.
Note the edge treatment on the cross to
the left. Ebendorf made the front of it
from the rusty inside of an olive-oil can
and bent its edges over so the painted
golden outside of the can becomes a
decorative frame, one with a contrasting
but harmonious color. The cross on the
right is made with the slightly discolored
outside of the can in front.

■ *Flying Bird Brooch (1994)*
Professor Robert Ebendorf
Silver, Mexican fire opal, broken English blue china plate, 18-karat gold bezel, and stone

■ *Parking Lot Art Necklace (1995)*
Professor Robert Ebendorf
Found objects from a parking lot, including many film cans

■ *Brooch: My Early Garden*
Roberta and David Williamson
This pin, a collaborative effort, is fabricated from sterling silver and includes a tin bird and a toy tin can.

■ *Necklace, earrings, and brooch: Hawaiian Nights*
Gege Kingston

■ *Necklace and earrings: Heard It on the Grapevine*
Gege Kingston

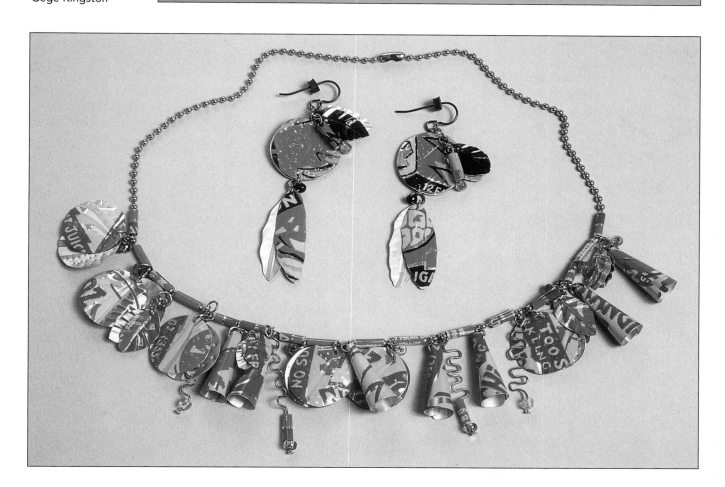

■ *Brooch: Fortune Series—Wait Til You've Paid for This One,* **Julie Flanigan**

JULIE FLANIGAN

The search for treasures amidst piles of junk is inspiring. Not knowing what you'll find (or what you're looking for) and how it could relate to a design always brings new challenges. Found materials have colors and surface textures that can't be created in a studio overnight. The most gratifying find might be a piece of scrap tin or steel with a painted graphic surface fading into a layer of rust. The materials have had another life and bring this former "personality" to the new creation. Juxtaposing common objects like a soup can with more "precious" materials adds humor to a design.

The objects I choose are primarily "man" made and weathered by Mother Nature before being recreated into sculptural designs.

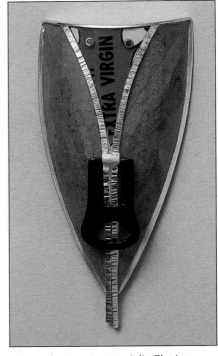

■ *Brooch: Easy Access,* **Julie Flanigan**

■ *Earrings: Fortune Series—Yes and No,* **Julie Flanigan**

■ *Cufflinks: Cheers,* **Julie Flanigan**

■ *Pin (1995),* Marjorie Simon
Sterling silver, Chinese turquoise, and
found object

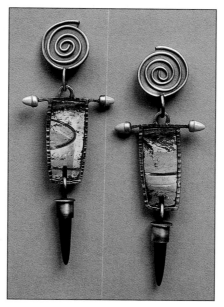

■ *Goya Guava Earrings (1995)*
Marjorie Simon
Sterling silver, Chinese turquoise,
frosted hematite, and Goya guava can

MARJORIE SIMON

The challenge of using found objects lies in transforming them to one's own voice. They already tell their own story and so provide a concert of narrative and surface. Sometimes the objects retain their recognizable form (rusted washers and heads on figurative pieces), and sometimes they're just another surface (cans with text).

Naturally I am aware of the political, economic, and emotional impact of using recycled materials, but I swear I choose them for their formal properties of texture, color, and form, and not for their sentimental value. Still, who can deny the richness of a rusted can or that as an artist one can give new life to the discarded? Though I occasionally stray into the school-lunch-box-memorabilia-school of contemporary jewelry making, I'm more concerned with using salvaged materials to provide a dark subtext out of the urban alluvia of my environment. (After all, I do live in New Jersey.) So I'll continue to walk the streets with my gaze in the gutter, plumbing the detritus of cyclone fences and parking lots; the most interesting stuff happens at the interstices of cultures anyway.

■ *Fix Fast Pin (1995),* Marjorie Simon
Sterling silver, found objects, frosted hematite,
and iron wire

■ *Johnson Wax Kit Earrings (1995),* Marjorie Simon
Sterling silver, amethyst, and frosted hematite

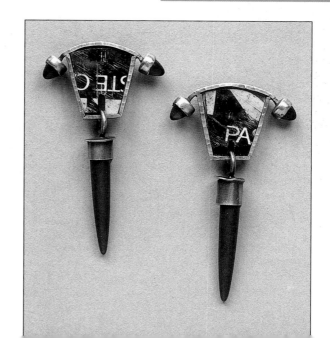

■ *Pins,* Shana Kroiz
A jeweler and metalworking professor, the artist constructed steel and plastic dies so she could use a hydraulic press to crush tin can sections into these sensual, rounded shapes. She fabricated the pins using time-honored jewelry methods.

■ *Flagship (1994),* Alistair Milne
Woven twisted wire and tin

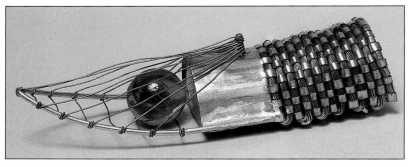

■ *Slipper (1994),* Alistair Milne
Twisted copper wire and tin

■ *Pin (1992),* Alistair Milne
Soldered tin and wire

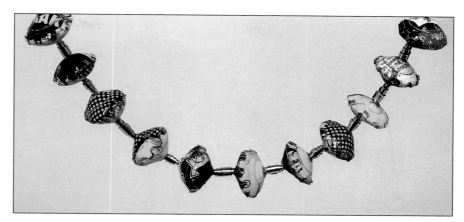

■ *Necklace Fragments*
Tim McCreight is a jeweler, a teacher, and the author of the best book on jewelry-making methods in the whole world (see "Further Reading" on page 144). His necklace fragments are textbook examples of how to work with tinned steel. Study them carefully. The large beads are made from film canisters. The end domes are made from the slightly altered canister caps. One set of beads contains discs cut from a phonograph record.

■ *Pendant* (left), **Lucille Beards**
Gilded can shard

■ *Pins,* **Jan Hutchinson**
Jan found a group of portraits painted on can lids in a flea market in Paris and collaborated by adding jewelry elements to create this intriguing group of pins.

■ *Brooch: The Importance of Little Things (1995)*
Keith E. Lo Bue
Front: Tin can lid, glass, photo, crab claws, beads, hardware, paper, and acrylic medium
Back: Mahogany, pin back, and paper

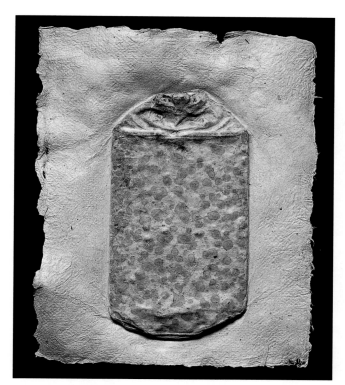

■ *Roadkill,* Betty Oliver
The artist found a flattened spray can in a parking lot. Back at her studio, she mixed up a batch of paper pulp and embedded the can on a sheet of handmade paper. Then she covered the can with another sheet to enclose the can completely. Rust from the can bled through the paper as it dried. The paper also shrank, leaving the can's image—as direct as a handprint and as elusive as a ghost.

■ *Buddha Fragment,* Betty Oliver
The can fragment pushing out from the paper evokes the sense of a continued presence, a "who" breathing there.

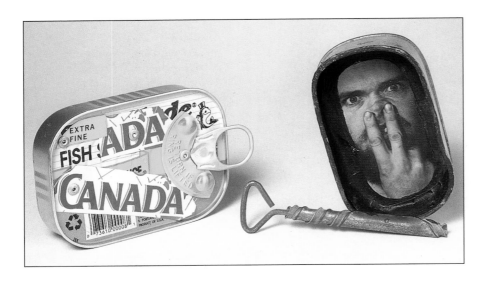

■ *Two Fish-Can Objects,* **Bobby Hansson**
My pal, Ada, is a Pisces, so I made the birthday present on the left to hold a photo of her. The can on the right is a self-portrait from the old days.

■ *Celestial Seasonings Necklaces (1994)*
Professor Robert Ebendorf
Coat-hanger wire, 18-karat gold, pearls, opal, and other stones

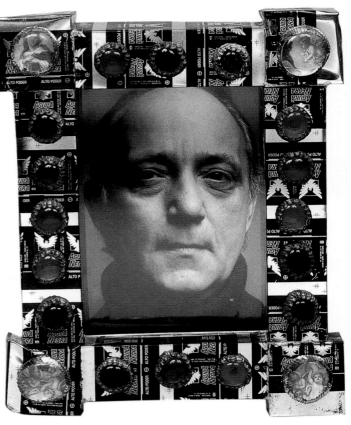

■ *Picture Frame,* **Lee Carter**

■ *Throw the Dog a Bone,* **John J. Grant**

■ *Three Men Go Fishing,* **John Grant**

■ *Bearing Gifts,* **John J. Grant**

■ Spoon Popkin
  Clockwise from upper left:
  Drinkers Head. **Notice the careful placement of the print in this amazing puppet head's eyes.**
  Drinking Armor (1988). **The artist models her drinking armor, which she made in Scotland.**
  Drinker's Hands (1988)

■ *Natty Boh Ship Headrest (1991),* **Dan Van Allen**
Dan models his own creation.

■ *Kali Niche (1991),* **Dan Van Allen**
The artist made this piece from a cookie can.

■ *Seahorse Cookie Cutter (1992),* **Spoon Popkin and Dan Van Allen**

■ *Cookie Cutters (1992),* **Spoon Popkin and Dan Van Allen**
As you can see, the skull cookies were shaped with two
of these cutters.

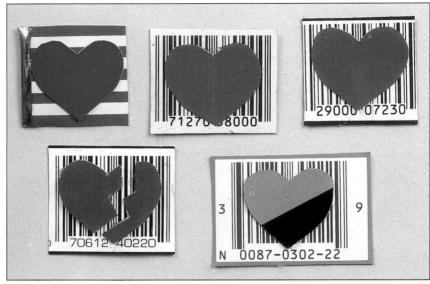

■ *Pins*
Flat Pin Studio
The Flat Pin Studio in Woodstock, New York, makes one-of-a-kind pins, each with a heart on a square background.

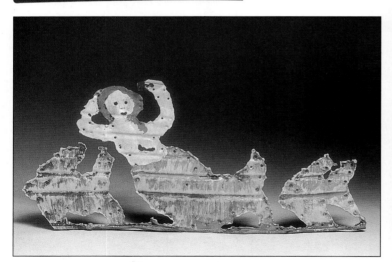

■ *Flattened Wall Pieces* (above and right), **Marcia Wilson**

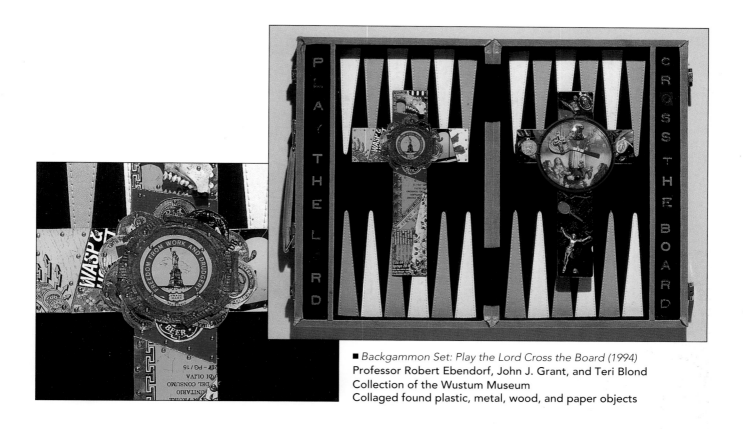

■ *Backgammon Set: Play the Lord Cross the Board (1994)*
Professor Robert Ebendorf, John J. Grant, and Teri Blond
Collection of the Wustum Museum
Collaged found plastic, metal, wood, and paper objects

■ *Pins,* Christine Kristen
The artist placed
images on can lids and
held them in place by
folding the edges of
the lids up and over
them. The corners were
embellished with beads
and found objects, and
findings were glued
onto the backs. One
pin has been photo-
graphed to show the
pin back.

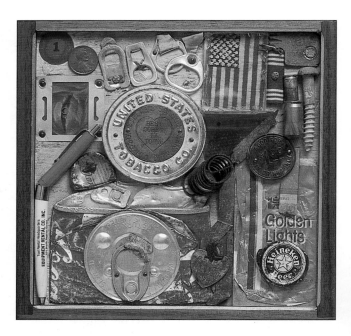

 *Wall hanging: The Late Great Disposable Society*

J. Fred Woell

Of the many pieces he's made from parts and pieces of tin cans, Fred explains that this one says the most about our wasteful society. When the switch on the side is turned on, a couple of LEDs flash every few seconds.

■ *Crushed Can Head with Tiddledywink Eyes* (above), **Judith Hoyt**

The nose and eyelids are cut and formed from sheet copper.

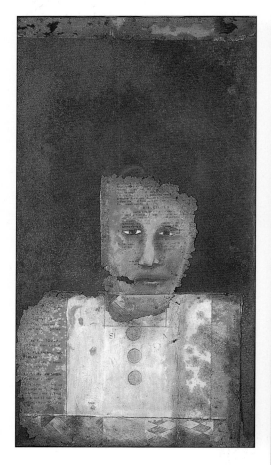

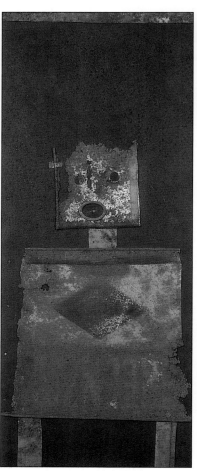

■ *Floor Wax Can Portrait* (far left)

Judith Hoyt

This bold construction of flattened cans is sensitively enhanced with paint to create an enigmatic portrait.

■ *Green and White Portrait* (left)

Judith Hoyt

The solitary, almost primitive figure seems very human in spite of its geometric composition.

■ *Holy Ghost Building #15,* Christina Shmigel
Based on houses along River Road, near New Roads, Louisiana

CHRISTINA SHMIGEL

Each of the pieces shown in these photos has a wood core to which the aluminum has been nailed. The cans were ones I found on walks, in the days before recycling came to Louisiana, or in abandoned camp-fires. I liked them for their baroque surfaces and for the faded traces of text and color. The structures are based on buildings that held for me a particular feeling of loss and nostalgia. The weathered, discarded cans seemed to serve this same end, although it is surprising to me that people rarely recognize the material for what it is.

Oddly enough, these pieces have become memorials, as many of the original buildings have been destroyed. At a recent slide lecture in Louisiana, the granddaughter of the owner of the Rayne rice dryer told me, to my great sadness, that it had fallen prey to arson.

■ *Holy Ghost Building #10,* Christina Shmigel
Based on a Spiritualist church in New Orleans, Louisiana

■ *Holy Ghost Building #22,* Christina Shmigel
Based on a rice dryer in Rayne, Louisiana

■ *Prisoner of Love (1967)*, Tony Berlant
14-5/8" x 10" x 14" (37.2 x 25.4 x 35.6 cm)
Found metal collage on plywood with steel brads,
polyester resin, plastic figure, and metal cage

■ *Aki (#130-1992)*, Tony Berlant
10-1/2" x 10" x 8" (26.7 x 25.4 x 20.3 cm)
Collage of found tin mounted on plywood with steel brads

■ *Cabin in the Sky (1984)*, Tony Berlant
22-1/4" x 18-1/2" (56.5 x 47 cm)
Found metal collage on plywood with brads in folk art frame

■ *Golden State (1963)*, Tony Berlant
Collection of Consie Miller
11-1/2" x 9-1/2" (29.2 x 24.2 cm)

■ *Indian Corn (1984),* **Tony Berlant**

■ *Dancing on the Brink of the World (1987),* **Tony Berlant**
San Francisco Airport Commission, 14' x 42' (4.2 m x 12.8 m)

TIN INTO GOLD

The dream of medieval alchemists was to turn base metal into gold. You may want to do the same thing with your tin can creations, but before you decide to sell your wares, you'll need to think seriously about some of the drawbacks of going commercial.

You'll need to come up with your own designs. On the one hand, learning new skills by copying other people's work is fine. Selling these copies, on the other hand, is not only unfair to the original artists, but may also land you in trouble with the law. You're welcome to copy my own designs—and even to sell these copies—but don't, under any circumstances, sell copies of anyone else's work without the artist's written permission.

Consider, too, the fact that you may be held liable for any injuries that your products cause. Unless you can afford insurance to protect you from suits of this sort, never sell anything that could cut someone, start a fire, or be swallowed.

The costs of working with tin may seem to be pretty low—after all, tin is free, right? If you plan to make a living income, however, you'll need to keep track of the cost of your tools and of the time you spend finding, washing, and handling tin cans, and marketing your work. A good way to get an idea of what your costs are—and to charge accordingly—is to spend at least one week recording every minute and penny you spend on your tin can creations. Divide that amount by the number of pieces you make during that time.

One final warning: You could succeed! You could very well find, as a friend of mine has, that your items sell so well that your lifestyle changes. Make sure that you're willing to make these sorts of changes before they hit you full force. There's a real difference between working at your own speed—and enjoying it—and cranking out handcrafted items in order to meet a deadline.

■ *Painted Tomato Can*
Michael Chelminski
Michael has a Masters degree in painting from Yale and usually paints on canvas. This is his first painting on a tin can.

■ *Kitchen Match Holders*
Bobby Hansson
I bought the antique lady match holder at an auction and made a gentleman friend for her out of an oil can.

CHAPTER TEN:
ART INSPIRED BY TIN CANS

■ *Is This Not a Can?*, M. Crenson
Age-old questions concerning reality and surreality are up in the air again when we contemplate this painting. Art history buffs will be familiar with a series of Magritte pictures of meerschaum pipes entitled *Ceci n'est pas une pipe*. Crenson's first name is Margaret, but perhaps this is coincidental.

IN PREVIOUS chapters, we've focused on tin cans as an amazingly versatile material—one that is used in hundreds of different ways. Now we'll sample some of the meanings cans have had for artists who find that tin makes an inspirational subject matter as well.

■ *Rolling Rock Basket,* John McQueen
Made on the spur of the moment as a gift for a weekend host, this whimsical beer-can basket grows from the intact round can bottom to a squared top rim.

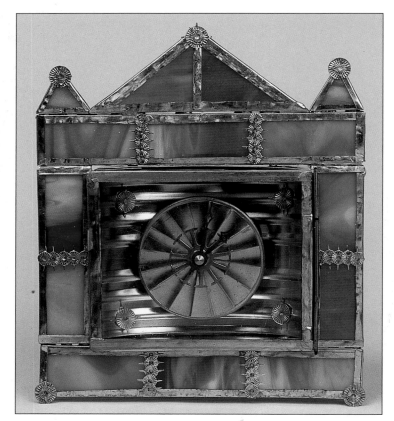

■ *Stained Glass Clock,* Randall Cleaver
The face of this clock and the channels holding the glass are made from tin cans.

■ *Matchbox,* Elizabeth Brim
Ms. Brim made the top of this container from a flattened Prince Albert can which was found on the highway. She fastened a tiger-eye stone to it and made the cast bronze oval box that it fits. Sulphur-tipped matches that are struck on the roughened surface contribute streaks of patina.

■ *Myself Divided and Dia De Muertos* (right and center)
Patrica Malarcher
The idea of "Plato's cave" reminds us that what humans perceive is an illusion, like the shadows cast on the wall of a cave by objects which exist outside the cave. These artworks are not tin cans, but are the shadows created when crumpled cans are scanned by the glaring, cold bright bar in a photocopy machine—shadows captured on paper and manipulated by the artist.

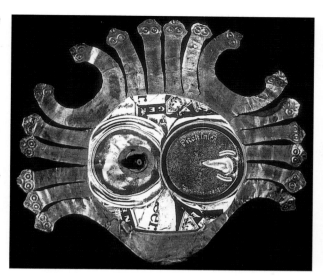

PATRICIA MALARCHER

I used to walk across a parking lot to get to my studio. Occasionally I would find a crushed can looking up at me from the blacktop, pleading for attention. Not every can had the capacity to do so, but those that did seemed to be asserting themselves as personages. Back in the studio, some of the cans suggested ways that their personalities could be more fully expressed. For some, it was only after they were photocopied that their hidden life came to light. In their crushed abandonment, they seemed to symbolize many of the unfortunate human conditions around us today.

One day, a Budweiser can, bent so that the two round ends came together, called out to me, and in it I saw an image of myself wearing large rimmed glasses. Around that time, I had been invited to participate in a book show and used photocopied images of this can for a sequential self portrait (Myself Divided) illustrating different aspects of my life.

Dia De Muertos is based on an image that showed up after I made a color copy of a can that I admired for the color and texture of its rust. I have re-photocopied it in black and white and have used it to create several other images.

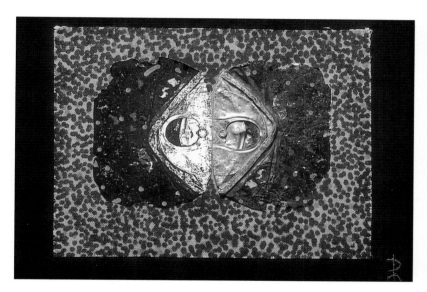

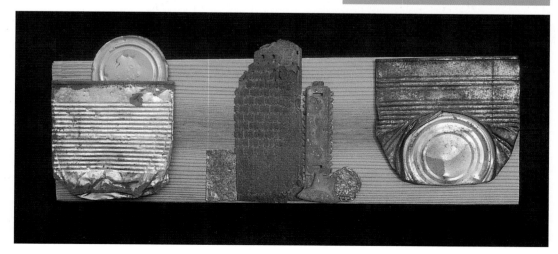

■ *Ancient City Scene with Sunrise and Sunset,* Billy Name
Photo by Dick Crenson
The artist made this construction from found objects.

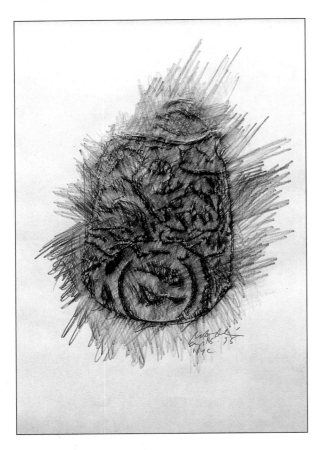

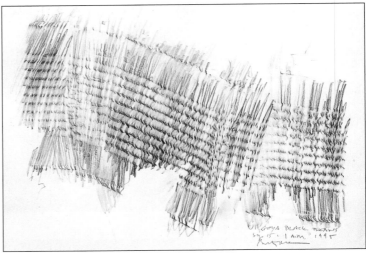

■ *Rubbing* (above)
Betty Oliver
An unopened can of black beans was rolled and rubbed at the same time to create this image. The dynamic progression of lines evokes the endless procession of cans rattling along a factory assembly line.

■ *Rubbing* (left)
Betty Oliver
The direction of the lines in the rubbing seem to suggest the forces which smashed the can flat.

■ *Hummingbird,* Elizabeth MacDonald
When Elizabeth moved to the country, her gentle cat became an intrepid hunter. Being a potter, Elizabeth thought it appropriate to cremate his offerings in closed containers when firing her kiln. She found that wet clay received the impressions of the victims' shapes and that the fired clay captured the fossilized forms. Bones, eggs, and wings focus her attention on the vastness of the universe. Images of Icarus and the phoenix rising give added meaning to these metaphors. An empty tin, which once held her small cigars, seems appropriate to contain this work.

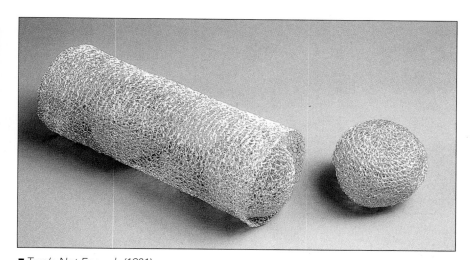

■ *Two's Not Enough* (1981)
Norma Minkowitz
Collection of Kenneth and Barbara Juster
Norma Minkowitz makes icons from everyday objects by transforming them into transparent crocheted sculptures which enclose and confine, but also expose. Structure and surface are achieved simultaneously, and depending on the light these icons are seen in, they can appear translucent or solid.

■ *Empty Words* (1982)
Norma Minkowitz
Collection of Carol Garvey

■ *Horn and Sheath,* Dan Mack
Dan makes rustic furniture, so he used a tree branch as a sheath to conceal and protect this horn, which, as you may notice, is made from a tin can turned inside out.

■ *Ceramic Pepsi Can Ashtray with Cigars*
Victor Spinski
When an artist takes an ordinary object like a smashed Pepsi can being used as a cigar ashtray, and sculpts it out of another material (in this case, clay with added glazes and ceramic photo decals), the whole reality is transformed.

■ *Orange Pekoe in a Can,* Robin Campo
Clay

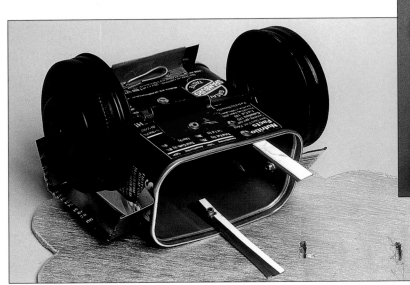

■ *Pam, Spam Jeepin' Down a Faux Bois Road*
Bobby Hansson
These luncheon meat cans make great jeeps as well as tanks. Studying the detail photo will help you understand how this sculptural portrait is constructed.

Chapter Eleven:
The Kikkoman Challenge

Clockwise from top left:

■ *Tin Briefcase,* Bobby Hansson

■ *Kikkocan on a Crutch,* Bobby Hansson
The notes of this instrument are determined by how hard you squeeze the can opener on the top; the opener tightens the string.

■ *Grand Opening Purse for Zette,* Bobby Hansson
Take a look at page 43 for another view of this purse.

■ *Soybean Watering Can,* Bobby Hansson
The author once lived in Decatur, Illinois—the "Soybean Capital of the World."

BY NOW, you'll be aware of my belief that tin cans are a wonderfully versatile raw material. Some of you will also have noticed that certain types of cans keep reappearing in this book. I don't work for Medaglia D'Oro or Hunt's, but their cans have pretty colors, and there are a lot of these cans in my area. I've enjoyed working with a limited palette. Seeing the same cans over and over again is like encountering old friends.

Kikkoman soy-sauce cans have always been a favorite of mine. They're big, flat, and include many different colors and patterns. As an experiment, I decided to send identical Kikkoman cans to ten different artists to see what each artist would create. I set only one rule: Whatever the artists made had to be returned within two weeks so I could photograph their creations for this book.

■ *Bird Brooch (1994)*
Professor Robert Ebendorf
Silver, 18-karat gold wire, pearls, and old tin can

■ *Pendant (1994)*
Professor Robert Ebendorf
Silver, 18-karat gold, pearls, and old tin can
Professor Robert Ebendorf is a fireball of energy and creativity. He sent me a postcard made from his can, telling me his work was on the way. Not long after, I received his two pieces, each adorned with pearls. Evidently the soy sauce reminded the professor of the Japanese pearl divers whom he admires.

■ *Kikkoman Poop Scoop*
Dan Van Allen
Dan Van Allen added an almost sacrilegious touch to the Kikkoman challenge by making this piece, which he uses to clean out his kitty litter box.

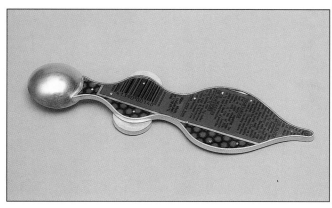

■ *Fabricated Silver Spoon,* **ROY**

ROY graciously interrupted her hectic schedule to create an intriguing and beautifully made spoon. On its back are some of the recipes printed on the can, and the front features magnificent bosomy shapes with the word "man" across them. Roy explains that this piece is about gender and duality. The implications of words and forms are left up to the viewer.

■ *Kikkoman Clock,* **Randall Cleaver**

Opening a package from clockmaker Randall Cleaver is always better than the best Christmas in the world. The note on the outside of the box said, "I hope this is what you had in mind. I got carried away." He sure did. The way in which Cleaver totally transformed the can and still gave the illusion of having left it intact is awesome.

When the clock doors are closed, the crystal doorknob legs, the brass-leafed cupola, and the massive copper doors are almost overpowering; the familiar beehive pattern is barely visible. But open the doors, and there's the label, cut out and pushed back, and then cut out and pushed back again. On the other side, the label is hinged to allow access to the clockworks and batteries, but otherwise the can seems untouched.

■ *Cans <u>Across</u> America: East Meets West*
Teri Blond
Mixed media

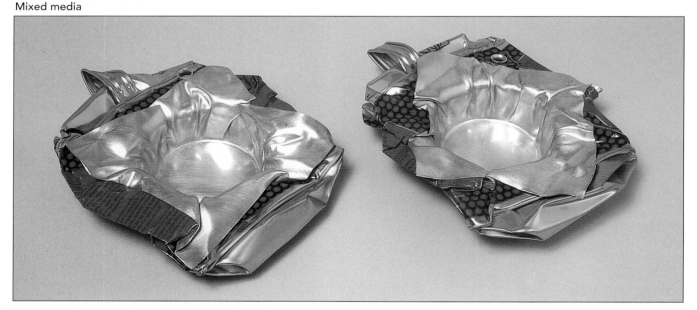

■ *Oh She Can Can* (left and above)
Teri Blond
Mixed media
Teri Blond, who has collaborated with Professor Ebendorf in the past (see page 123), also sent two works—wonderfully ornate and lush extravaganzas that dazzle the eye as they amaze the mind.

■ *Compacted Kikkoman Bowls,* Boris Bally

Boris Bally is an internationally known metalsmith who is famous for the big bowls he makes out of recycled aluminum traffic signs. When I called to invite him to participate, he immediately agreed, but added, "I use a 25-ton press, and I'm not used to working with tin cans, so please send me two of them, in case I ruin one."

As it turned out, Boris liked his first tin project—a bowl—so much that he decided to make a pair of them. He placed a sheet of lead-free pewter on top of each can and squashed it into a bowl-shaped die mold. Then he punched holes and installed aluminum grommets so the bowls can be displayed on a wall when they're not in use.

At first glance, these bowls seem to be identical, but if you study them carefully, you'll see the subtle differences created by the immense force of Bally's press.

■ *Can Tickle for Bobby*
Bennett Bean
Ceramic artist Bennett Bean said he needed a snapshot of me before he could complete his work. I was a little nervous about what Bean had in mind, but my wife said, "Fair is fair," and got out the portrait lens for her camera.

Bean had a twinkle in his eye when he pointed out that right next to my mug shot, the words "Oman Sauce" appear. Was he alluding to my enthusiasm for strong drink and rich food in the old days when he and I were younger and wilder?

BENNETT BEAN

If we'd done this 40 years ago, we could have combed our own local dump and had a great variety of cans to choose from, but recycling and the switch to paper labels on most cans in this country lead us to cans imported from countries that still use printing rather than paper. It's the globalization of can art.

I had never purchased cans of food based solely on the aesthetics of the labels, but this project required eating a number of things that I wouldn't have chosen otherwise. My favorite can contained preserved squid. After claiming the can, I offered the squid to my wife, who is Chinese, but racial solidarity was insufficient reason for her to eat them. They're still in the refrigerator, a monument to our neurosis about throwing away food, as well as to Hansson's obsession with can art.

■ *Envelopes with Hinged Flaps, Wall Sconce, and Candle Extinguisher*
Richard Haddick
Tinsmith Richard Haddick took one look at the Kikkoman can and reached for his shears. He cut the top and bottom off and flattened the can out, announcing, "I see a pair of wall sconces."

He designed a sconce reflector to fit the logo on the can and used the beehive pattern for the arm which holds the candle, folding the metal so the pattern would show from both sides. The long-handled candle extinguisher was the finishing touch.

"I'd like to send the company a note thanking them for making such beautiful cans. Maybe they'll send me more," he mused, and made a tin envelope with a hinged flap.

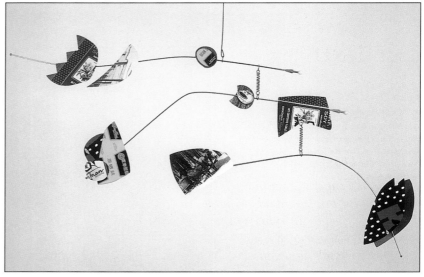

■ *Kikkomobiles,* Jean-Pierre and Carol Hsu
Carol and Jean-Pierre Hsu collaborated and sent two Kikkomobiles. One has arms made from recycled automobile antennae. It swoops and swings every time we open the door and sends little light beams all around the room when the sun hits it. The other one is smaller, with curled arms that look almost like calligraphy lines. This one spins in lazy circles on its own center. It's amazing to see how different two mobiles, both made from the same tin can, can be.

■ *The Knokkoff Can Toy Soy Truck*
Harvey Crabclaw
Harvey used part of a Kikkoman can in his Coleman frame (see page 58), so I didn't even think of sending him one of the cans. He heard about the Kikkoman challenge anyway and sent a package along with a note that said, "Kikko cans nowhere to be found in New Mexico, but how about this?"

Inside the box was a Monggojin soy-sauce can, almost identical in color and design to the Kikkoman can. Harvey had used the copycat can to make a copy of one of my trucks. He also copied my idea by placing the real Elvis in the driver's seat. I'm sincerely flattered. Thanks, pal o' mine, to quote Ralphie Boy.

Chapter Twelve:
The End

Dear Bobby,

I was in a supermarket in Blacksburg, Virginia, thinking about your book, and what I might say about tin cans, when I saw a little boy carrying a big can of tomatoes for his father, who had just rounded the corner into the next section. Finding himself alone, the boy set the can on its side and used his foot to roll it the rest of the way down the aisle. When he reached the end of the aisle, he picked up the can and disappeared around the corner.

How could I express our nearly worldwide impulse to create any better than this little boy's spontaneous gesture of invention? From his hand to the floor, from the floor to his foot—in those instants, a can became a wheel.

Later, I caught up with him in another aisle and said, "I saw you rolling that can." He looked a little scared, as if he thought I might scold him, so I quickly added, "Maybe I should follow you around,"—and now he really looked scared—"to see what else you invent. You see, a friend of mine is writing a book about the things you can make with tin cans, and I was so interested to see you use that one as a wheel!"

I could almost see his cerebral wheels turning as he sized me up. He thought a moment, then began, his words almost spilling over each other. "I know what you could do. You could take the two small cans and put them side by side like this for wheels, and then take a larger one and lay it across the others like this and make a tank, or you could get five cans, three large and two small, and put them like this, and…"

He really said it all.

Love,

Betty Oliver

■ *The End,* Betty Oliver
The title says it all. And why flattened rusty cans? As images, they resonate throughout the world culture and across class barriers to almost every consumer. Tin cans reflect us back to ourselves as this piece reflects the can, encouraging us toward another way of viewing the world.

CONTRIBUTING ARTISTS AND COLLECTORS

The author: **Bobby Hansson**
Rising Sun, MD
Pages 3, 8, 9, 15, 18, 19, 20, 21, 22, 24, 25, 30, 33, 35, 37, 38, 39, 40, 42, 43, 44, 47, 49, 63, 73, 74, 80, 81, 83, 84, 85, 88, 89, 90, 92, 94, 97, 98, 99, 100, 101, 102, 106, 118, 128, 134, 135, and cover

Christopher Anna
New York, NY
Page 40

W.H. Bailey and Laurie B. Eichengreen
New York, NY
Pages 51, 78, 80, 86, 87, and 90

Boris Bally
Pittsburgh, PA
Page 138

Bennett Bean
Blairstown, NJ
Page 139

Lucille Beards
Havre de Grace, MD
Pages 25 and 116

Tony Berlant
Santa Monica, CA
Pages 7, 107, 126, and 127
Photos courtesy of the artist

Harriette Estel Berman
San Mateo, CA
Pages 52 and 53

Teri Blond
San Antonio, TX
Pages 109, 123, and 138

Joseph Bomba
Port St. Lucie, FL
Page 91

Ricki Boscarino
Montague, NJ
Page 41

Elizabeth Brim
Penland, NC
Pages 24 and 130

Robin Campo
Chamblee, GA
Page 134

Lee Carter
San Francisco, CA
Pages 40, 69, 81, and 118

Michael Chelminski
Bridgewater, CT
Page 128

Tina Chisena
Wheaton, MD
Page 32

Randall Cleaver
Landsdowne, PA
Pages 56, 130, and 137

Janet Cooper
Bottlecap Jewelry
Sheffield, MA
Pages 31, 45, 54, and 93

Harvey Crabclaw
Santa Fe, NM
Pages 13, 57, 58, 59, 60, 61, and 140

Dick Crenson
Pleasant Valley, NY
Pages 38, 44, 49, and 69
Photos courtesy of the artist

Margaret Crenson
Pleasant Valley, NY
Page 129

Lynda Curtis
New York, NY
Page 92

Robert Dancik
South Salem, NY
Page 31

John Daniels
Stafford, VA
Page 47

Chris Darway
Lambertsville, NJ
Page 27

Jose de Creeft
New York, NY
Page 10

Sean Duffy
Los Angeles, CA
Page 54

Daniel Eaves
Mechanicsville, VA
Pages 17 and 23

Professor Robert Ebendorf
Rosendale, NY
Pages 109, 110, 118, 123, 136, and cover

Sue Eyet
Port Deposit, MD
Pages 27, 34, 46, 50, 55, 93, 98, and cover

Ruth Abraham Fisher
London, England
Pages 54 and 108

Julie Flanigan
Baltimore, MD
Pages 93 and 112

Laurie Flannery
Baltimore, MD
Page 70 and back cover

Flat Pin Studio
Woodstock, NY
Page 122

John J. Grant
Santa Monica, CA
Pages 5, 11, 119, 123, and cover

Reneé Habert and Jim Stonebraker
Hoboken, New Jersey
Page 30

Richard Haddick
Wyoming, DE
Pages 12, 24, 62, 63, 64, 65, 71, 75, 76, 77, 82, 139, and cover

Edmundo de Marchena Hernandez
New York, NY
Page 32

Annabel Hill
London, England
Page 96
Photos courtesy of the artist

Homefront
Hillsborough, NC
Page 79

Ruthann and Jerry Hovanec
Lusby, MD
Page 22

Judith Hoyt
New Paltz, NY
Page 124

Carol and Jean-Pierre Hsu
Berkeley Springs, WV
Page 140

Rob Hudson
Rock Hall, MD
Page 26

Jan Hutchinson
Pass Christian, MS
Page 116

Gege Kingston
Stockbridge, MA
Page 111

Christine Kristen
San Francisco, CA
Page 123

Shana Kroiz
Baltimore, MD
Page 114

Jody Kruskal
Brooklyn, NY
Pages 103, 104, and 105

Karen Lasley
Klassic Spirits
Paducah, KY
Page 22

Pamela Lins
Brooklyn, NY
Page 76

Keith Lo Bue
Westport, CT
Page 117

METRIC CONVERSION CHARTS

Linear Measurements

Inches	CM	Inches	CM
1/8	0.3	20	50.8
1/4	0.6	21	53.3
3/8	1.0	22	55.9
1/2	1.3	23	58.4
5/8	1.6	24	61.0
3/4	1.9	25	63.5
7/8	2.2	26	66.0
1	2.5	27	68.6
1-1/4	3.2	28	71.1
1-1/2	3.8	29	73.7
1-3/4	4.4	30	76.2
2	5.1	31	78.7
2-1/2	6.4	32	81.3
3	7.6	33	83.8
3-1/2	8.9	34	86.4
4	10.2	35	88.9
4-1/2	11.4	36	91.4
5	12.7	37	94.0
6	15.2	38	96.5
7	17.8	39	99.1
8	20.3	40	101.6
9	22.9	41	104.1
10	25.4	42	106.7
11	27.9	43	109.2
12	30.5	44	111.8
13	33.0	45	114.3
14	35.6	46	116.8
15	38.1	47	119.4
16	40.6	48	121.9
17	43.2	49	124.5
18	45.7	50	127.0
19	48.3		

Volumes

1 fluid ounce	29.6 ml
1 pint	473 ml
1 quart	946 ml
1 gallon (128 fl. oz.)	3.785 l

Weights

0.035 ounces	1 gram
1 ounce	28.35 grams
1 pound	453.6 grams

Sources

Further Reading

The Complete Metalsmith *by Tim McCreight*
The Brynmorgen Press
33 Woodland Road
Tel: (207) 767-6059
Fax: (207) 799-1172

Although this book doesn't include much information on tin cans, it's an invaluable source of information about tools and techniques. Brynmorgen also carries other good metalworking books.

Video

Dale the Tinker
P.O. Box 21
St. Albans
West Virginia 25177

Made at the Fort New Salem Museum, this video, which comes with a booklet, shows a colorful old-time tinsmith at work. Dale uses traditional techniques—with some funky homemade improvements—to create historic items.

Audiotapes and CDs

Bobby Hansson
P.O. Box 1100
Rising Sun, MD 21911

Write to the author for a catalogue of tapes and CDs of tin-can musical instruments.

Suppliers

M.S.C. Industrial Supply Co.
3051-A Washington Blvd.
Baltimore, MD 21230
Tel: (800) 934-3008

M.S.C., which has branches across the United States, offers tin snips, pliers, hammers, rivets, letter stamps, soldering tools and supplies, gloves, goggles, and paint—in short, everything you will ever need or want, including an 800 number.

Precision Movements
4283 Chestnut Street
P.O. Box 689
Emmaus, PA 18049
Tel: (800) 533-2024

Clockworks and parts of every description are available through Precision Movements.